Transformation Toughening of Ceramics

Authors

D. J. Green, Ph.D.
Department of Materials Science and Engineering
The Pennsylvania State University
University Park, Pennsylvania

R. H. J. Hannink, Ph.D., and M. V. Swain, Ph.D.
Division of Materials Science and Technology
Commonwealth Scientific and Industrial Research Organization
Clayton, Victoria, Australia

CRC Press
Taylor & Francis Group
Boca Raton London New York

CRC Press is an imprint of the
Taylor & Francis Group, an **informa** business

First published 1989 by CRC Press
Taylor & Francis Group
6000 Broken Sound Parkway NW, Suite 300
Boca Raton, FL 33487-2742

Reissued 2018 by CRC Press

© 1989 by CRC Press, Inc.
CRC Press is an imprint of Taylor & Francis Group, an Informa business

No claim to original U.S. Government works

Library of Congress Cataloging-in-Publication Data

Green, D. J.
    Transformation toughening of ceramics.

    Bibliography: p.
    Includes index.
    1. Ceramic materials—Mechanical properties.
I. Hannink, R. H. J.   II. Swain, M. V.   III. Title.
TA455.C43G74   1989   666   88-25805
ISBN 0-8493-6594-5

A Library of Congress record exists under LC control number: 88025805

Publisher's Note
The publisher has gone to great lengths to ensure the quality of this reprint but points out that some imperfections in the original copies may be apparent.

Disclaimer
The publisher has made every effort to trace copyright holders and welcomes correspondence from those they have been unable to contact.

ISBN 13: 978-1-315-89830-8 (hbk)
ISBN 13: 978-1-351-07740-8 (ebk)

Visit the Taylor & Francis Web site at http://www.taylorandfrancis.com and the
CRC Press Web site at http://www.crcpress.com

PREFACE

The aim of this book is to provide a coherent and up-to-date discussion of the scientific work concerning the transformation toughening of ceramics. We hope the book is useful to scientists, engineers, and students who are new to these materials. It is intended both as a source of learning and information to those who are actively involved in studying the mechanical behavior and microstructural relationships in transformation-toughened ceramics. While it has been our aim to present a book that is as current as possible at the time of publication, the subject is still expanding in many areas; so our main hope is that the reader will also gain an insight into the direction of future advances.

The authors would like to acknowledge the contributions, discussions, and inspiration from our many colleagues and friends who have been concerned with the development of this field of Materials Science. In particular, we would like to acknowledge the critical contributions and encouragement of Professors Pat Nicholson, Fred Lange, Nils Claussen, Arthur Heuer, and Tony Evans. In Australia we would like to credit the inspiration and guidance of the late Dr. N. A. McKinnon, who was responsible for initiating and nurturing the Zirconia Project at CSIRO in the late sixties. It was Neil who was responsible for recruiting the team that developed transformation toughening and knew something exciting would develop from this work. Discussions and work, in the early stages of the Zirconia Project, with the late Drs. R. T. Pascoe and K. A. Johnston were also a pleasure and fruitful for RHJH in particular. We are grateful for the inspiration, excellent technical support and stimulating discussions with our current and ex-colleagues; J. Allpress, J. Drennan, R. C. Garvie, V. Gross, R. R. Hughan, L. R. F. Rose, H. R. Rossell, H. G. Scott, and R. K. Stringer at CSIRO, D. R. Clarke, B. I. Davis, M. G. Metcalf, P. E. D. Morgan, T. M. Shaw, and E. Wright at Rockwell International Science Center. We also acknowledge the work of Jill Glass in critically reading the final version of the book and D. B. Marshall for his comments on Chapter 3.

For all of us (DJG, RHJH, and MVS), it has been our luck that we were involved with the development of a new class of materials and the excitement this has entailed. Finally, the effort needed to complete the book would have been impossible without the tolerance and support of our wives, Chris, Jill, and Helen, respectively.

THE AUTHORS

David J. Green, Ph.D. — After graduating from the University of Liverpool, England with a B. Sc. (Chemistry/Materials Science) in 1970, he completed his graduate studies in Materials Science at McMaster University, Canada. In these studies, he demonstrated the importance of microcracking in the fracture of ZrO_2-based materials and utilized ultrasonic fractography for studying crack-particle interactions in brittle fracture. In 1975, Dr. Green joined the Canadian Federal Government to work in the Department of Energy, Mines, and Resources. The primary emphasis of this work was concerned with the preparation of ultrafine, homogeneous ceramic powders for the fabrication of solid electrolytes. Joining Rockwell International Science Center in 1979, Dr. Green continued to study the relation between fabrication, microstructure, and the properties of ceramic materials. This research included microcracking in ceramics, reliability of ceramics in structural design, failure analysis, fabrication and evaluation of transformation-toughened ceramics, surface stresses in ceramics, and the mechanical behavior of lightweight ceramics. In 1984, Dr. Green joined The Pennsylvania State University as an Associate Professor and is combining his research on the mechanical behavior of ceramics with teaching. Dr. Green is a member of the American and Canadian Ceramic Societies and the Materials Research Society.

Richard Hannink, Ph.D., studied metallurgy at Newcastle Technical College, Australia, and the Sir John Cass College, London, after which he was awarded an A.I.M. in 1970. In 1973 he received a Ph.D. degree in physics from Cambridge University. He has worked at John Lysaghts and Richard (Australia) and Thomas and Baldwyns (England) on electrical and deep drawing properties of steels. His Ph.D. studies involved aspects of deformation in cubic transition metal carbides. In 1973 he joined the Commonwealth Scientific and Industrial Reasearch Organization (CSIRO) in Australia where he became codiscoverer of transformation toughening in ceramic materials. His research interests remain involved with projects aimed at improving the strength, toughness, and wear resistance of zirconia and zirconia-toughened composites, and development and application of wear-resistant materials in general. He is particularly interested in the correlation of microstructures with mechanical properties and *in situ* performance of such materials. He is currently leader of the Structural Ceramics Group at CSIRO, a member of the Australian Ceramic Society, Materials Research Society (U.S.), and a member of the editorial board of *Materials Forum*, the technical journal of the Institute of Metals and Materials Australasia.

Michael Vincent Swain, Ph.D., completed his undergraduate studies in physics at the University of New South Wales in Sydney in 1968. He then continued on at the same university to complete a Ph.D. under the supervision of Dr. Brian Lawn on "Some Aspects of Brittle Fracture in Intrinsically Strong Solids". Since completing his Ph.D., he has worked in various corporate, governmental, and academic institutions in the U.S., U.K., France, Germany, and Japan. For the last 10 years he has been with Commonwealth Scientific and Industrial Research Organization (CSIRO) in Australia and is currently leader of the Ceramic Composites Group. His research interests are primarily in the fields of mechanical properties of ceramics, particularly the role of microstructure on properties. He also has interests in the fields of glass, rock mechanics, as well as indentation deformation and fracture mechanics. More recently his interests have been in the areas of toughened ceramics and fatigue behavior of such materials. He is a member of the Australian and American Ceramic Societies, the Institute of Metals and Materials Australasia, and the Materials Research Society.

GLOSSARY OF SOME ZIRCONIA TERMS

A_f	Temperature for finish of martensitic transformation (during heating)
A_s	Temperature for start of martensitic transformation (during heating)
Ca-CSZ	Calcia cubic-stabilized zirconia
Ca-PSZ	Calcia partially stabilized zirconia
Ce-TZP	Ceria tetragonal zirconia polycrystal
CSZ	Cubic (fully) stabilized zirconia
$c-ZrO_2$	Cubic zirconia
HIP	Hot isostatic pressing (or pressed)
K_{IC}	Fracture toughness (or critical stress intensity factor)
LSM	Localized soft mode
Mg-CSZ	Magnesia cubic-stabilized zirconia
Mg-PSZ	Magnesia partially stabilized zirconia
M_f	Temperature for finish of martensitic transformation (during cooling)
M_s	Temperature for start of martensitic transformation (during cooling)
$m-ZrO_2$	Monoclinic zirconia
MOR	Modulus of rupture (more recently, flexural strength)
OA	Overaged
PA	Peak aged
PSZ	Partially stabilized zirconia
SPG	Secondary precipitate growth
TTZ	Transformation-toughened zirconia
TZP	Tetragonal zirconia polycrystal
$t'-ZrO_2$	Nontransformable tetragonal zirconia
$t-ZrO_2$	Tetragonal zirconia
UA	Underaged
Y-PSZ	Yttria partially stabilized zirconia
Y-TZP	Yttria tetragonal zirconia polycrystal
ZDC	Zirconia dispersed ceramics
ZTA	Zirconia-toughened alumina
ZTC	Zirconia-toughened ceramics
ZTM	Zirconia-toughened mullite
ZTS	Zirconia-toughened spinel

TABLE OF CONTENTS

Chapter 4

MICROSTRUCTURE-MECHANICAL BEHAVIOR OF PARTIALLY STABILIZED ZIRCONIA (PSZ) MATERIALS

Chapter 5

MICROSTRUCTURE-MECHANICAL BEHAVIOR OF ZIRCONIA-DISPERSED CERAMICS (ZDC)

Chapter 6
SURFACE MODIFICATION OF ZIRCONIA-TOUGHENED CERAMICS (ZTC)

Chapter 1

INTRODUCTION

I. PURPOSE

In recent years, there has been a renaissance in the science and technology of ceramics. This renewed interest has been spurred by the versatility in the properties of these materials and the myriad uses in which these properties can be exploited. Traditionally, the applications of these materials emphasized either high temperature applications (refractories) or inexpensive consumer items (bricks, whiteware, etc.). It is now realized that ceramics have a special set of electromagnetic, thermal, electrical, chemical, magnetic, and mechanical properties that allows them to be used in many ways.[1] The emphasis in this book is directed to the mechanical behavior of ceramics and, in particular, on a phenomenon known as transformation toughening. The basic and applied research on this effect has produced a new generation of tough and strong ceramics, which are currently being exploited in a wide variety of applications. It is, therefore, important that both the materials scientists and engineers understand the scientific principles on which these materials are based. Fortunately, many of the phenomena involved in transformation toughening are relatively well understood and the science has been well developed. Thus, the aim of this book is to review the current understanding of transformation toughening in a relatively basic way, so that the book is not only a useful source of information but also can be used to learn the principles involved in these materials.

II. CERAMICS IN STRUCTURAL APPLICATIONS

The current interest in the mechanical behavior of ceramics is often related to their use as a structural material. Ceramics and other inorganic materials have been used as structural materials for millennia, but invariably as compressive members.[2] This approach was taken because the compressive strength of ceramics is generally considerably higher than the tensile strength. In most modern engineering structures, however, tensile members are used extensively and the use of inorganic, nonmetallic materials has become much less common. The dominating feature of the tensile behavior of ceramics is their susceptibility to brittle failure. It is important here to distinguish the difference between brittleness and strength. Brittleness implies it does not take much energy to propagate a crack through a material, whereas strength is the stress required to initiate the propagation of the crack through the material. Indeed, as we shall see, the strength of ceramics can be extremely high, but once a crack starts to propagate, the failure is often immediate and catastrophic. In recent years, however, several approaches have been taken and have relaxed the concept that "ceramics are inherently brittle". These approaches include the incorporation of ceramic fibers or metal particles into ceramic matrices and transformation toughening. These toughening mechanisms have started to bridge the gap between the brittleness of ceramics and the ductility of metals.

Transformation toughening is based on the idea that a phase transformation can be stress induced in a material in such a way that it decreases the driving force that is acting to propagate the cracks that are present, or can form under stress, in the material. There are strict ideas about the type of phase transformations that can be utilized, but one would expect that several materials could be used in such a way. At the moment, however, there has only been one material, zirconia (ZrO_2), that has been successfully and extensively exploited as the transformation toughening agent. The remaining part of this chapter will, therefore, be used to review the science and technology of zirconia. This will be done from both a historical perspective and a materials science viewpoint. This latter approach stresses the relationship between the structure of a material, i.e., atomic, electronic, microscopic, etc., the properties of a material, and the fabrication procedure that is used to produce a specific structure. The interplay between these elements represents a basic philosophy in materials science. There have been previous and extensive reviews on the science and technology of zirconia and these should be consulted for further details.[3-6]

III. HISTORICAL PERSPECTIVE

ZrO_2 was discovered in Brazil as the naturally occurring mineral, baddeleyite, by Hussak in 1892.[3] This rich mineral usually contains about 80% ZrO_2, but it can be as high as 90%.[3] The major impurities in this mineral are usually TiO_2, SiO_2, Fe_2O_3, etc. The other main source of ZrO_2 is zircon ($ZrSiO_4$), which occurs as secondary deposits in India, Australia, and the U.S. ZrO_2 is not a rare substance and represents about 0.02 to 0.03% of the Earth's crust. It is, however, often found in small concentrations. For example, $ZrSiO_4$ is found in igneous rocks and crystalline schists and the crystals possess a high refractive index and are often prized as gems. Zircon crystals can be water clear, somewhat like diamonds, or can display vivid colors such as the topaz-colored hyacinth stones of Sri Lanka. Thus, although ZrO_2 is relatively abundant, rich ZrO_2-yielding ores are relatively scarce, but less so than many metallic ores, such as tin or copper.

Zr belongs in the fourth group of the Periodic Table, between Ti and Hf. The chemistry and crystal structures of HfO_2 and ZrO_2 are very similar. Moreover, HfO_2 is often found in ZrO_2 ores with the amount varying from 2 to 22%. The first industrial use of ZrO_2 was as a refractory material by Germany in World War I, but as we shall see, there are many unusual properties of this material in addition to its refractoriness.

Pure ZrO_2 exists in three different crystal structures, i.e., monoclinic, tetragonal, and cubic. The monoclinic crystal structure was identified by Ruff and Ebert[7] using X-ray diffraction. In addition, they found that the monoclinic form undergoes a disruptive phase transformation to the tetragonal form ~1100°C. The phase transformation usually leads to the shattering of a body, but it was found that ZrO_2 could form a fluorite-type cubic phase by the addition of various metal oxides.[8,9] This process removes the disruptive transformation and stabilizes the cubic phase at low temperatures and, hence, such materials were called (cubic) stabilized zirconias (CSZ).

The alloying oxides used to stabilize the cubic phase act to lower the phase transformation temperatures as they are added to the ZrO_2. Thus, it is possible to produce materials which are a mixture of the cubic and monoclinic (or tetragonal) phases. These materials are called partially stabilized zirconias (PSZ) and were found to be useful because they possess better thermal shock resistance than CSZ. The equilibrium stability regions of phases are generally shown in phase diagrams. Figure 1 shows an example of possible zirconia-metal oxide diagram in which the solid solution cubic phase is formed. In this diagram, the monoclonic/tetragonal transformation in pure zirconia is shown to occur at 1200°C. As we add metal oxide, we find solid solution regions for the tetragonal and monoclinic phases and the transformation temperature decreases to <600°C at the eutectoid composition of ~4.8 mol% metal oxide. For larger metal oxide additions, the solid solution cubic phase appears, in which the cubic phase is stable down to room temperature. The single phase, solid solution regions are separated by various two-phase regions, i.e, cubic + tetragonal, cubic + monoclinic, and tetragonal + monoclinic. The phase diagrams of the binary oxide systems that form CSZ and PSZ have been reviewed by Stubican and Hellmann.[10]

The tetragonal-monoclinic phase transformation in ZrO_2 has been the subject of considerable study. Wolten[11] was the first to suggest that the transformation was martensitic, i.e., similar to that of martensite, which is used to harden steels by specific heat treatments.

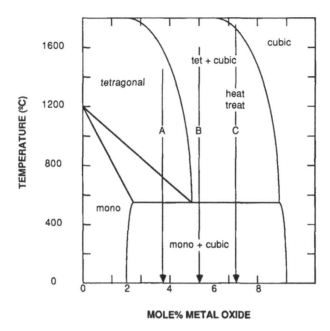

FIGURE 1. A schematic metal oxide-zirconia phase diagram showing routes for fabricating three different microstructures.

Similarly, it was the studies of the ZrO_2 phase transformation and how it could be controlled by heat treatments in PSZ, that led to the discovery of transformation toughening.[12]

As discussed above, it had been found that the PSZ materials often had superior mechanical properties compared to the CSZ materials, particularly their thermal shock resistance. This led to a variety of scientific studies that were aimed at understanding the mechanical behavior of PSZ. It was clear that the lower values of the thermal expansion coefficients compared to CSZ were part of the answer, but the attention turned to the microstructure of the PSZ materials in order to understand the effect of the tetragonal to monoclinic phase transformation on the mechanical properties. King and Yavorsky[13] argued that the stresses that accompany the transformation led to plastic deformation of the cubic grains and this form of stress relief reduced the possibility of failure. In a later study by Garvie and Nicholson,[14] it was suggested that microcracking, rather than plastic deformation, was the cause of the superior thermal shock resistance, that is, the stresses produced by the phase transformation were relieved by the formation of microscopic cracks. These microcracks were associated with the local stress field of the transforming particles and, hence, could be arrested. It was then suggested using a concept put forward by Hasselman,[15] that such microcracks would propagate quasi-statically in response to thermal stresses. In other words, if there is a sufficient density of microcracks, they propagate in a stable fashion so that rather than a sudden reduction in strength, thermal shock would lead to only a gradual reduction in strength.

At this point, it was realized that the microstructure of PSZ materials depended on the details of the fabrication process. These materials were invariably sintered in the cubic phase field, but then on cooling they passed through the tetragonal-plus-cubic field, in which the tetragonal phase precipitates, and then finally the tetragonal phase undergoes the phase transformation to monoclinic. The evolution of intergranular precipates discussed by Green et al.,[16] for a Ca-PSZ, and Bansal and Heuer,[17] who were studying a Mg-PSZ, showed that in addition to the intergranular precipitates, there was also a finer dispersion of precipitates within the grains (intragranular). The idea was then put forward that the microstructure could be manipulated by heat treatments in which the precipitation of the tetragonal phase was controlled.[18] This turned out to be a key philosophy because as we shall see shortly, for particular heat treatments, the tetragonal phase precipitates but does **not** transform to monoclinic on further cooling. The retention of the high temperature tetragonal phase to room temperature is the key to the fabrication of transformation-toughened materials.

At the same time as the work on thermal shock resistance and microstructural evolution of PSZ was underway, studies were initiated to look at the strength and fracture toughness of these materials. The fracture toughness of a Ca-PSZ was measured by Green et al.[19] and it was found in this particular material that microcracking had occurred during fabrication and though the material had a low strength, it possessed a high fracture surface energy. It was suggested that the propagation of a crack in this material involved the presence of a microcracked process zone at the crack tip and that the energy absorbed in this zone was responsible for the toughness. For a Mg-PSZ, Bansal and Heuer[20] suggested that changes in toughness were a result of the cracks being impeded by the fine dispersion of monoclinic precipitates within the cubic grains.

The discussion in this section has considered materials in which tetragonal (t-) ZrO_2 precipitates and then later transforms to monoclinic (m-) ZrO_2. It was, however, realized by Claussen[21] that ZrO_2 could also be introduced as a second phase by producing particulate composites. In a study on the mechanical behavior of Al_2O_3-ZrO_2 composites, he found that the ZrO_2 increased fracture toughness and attributed this to stress-induced microcracking. It is interesting to note that a major industry based on the use of fusion cast Al_2O_3-ZrO_2 as an abrasive in grinding wheels had been developed previously and may have indicated that such materials have useful mechanical properties.

The major breakthrough in the mechanical behavior of PSZ came from the work of Garvie et al.,[12] when it was realized that toughening could be produced in materials which contained substantial amounts of the metastable, intragranular tetragonal phase, and that the toughening was associated with the stress-induced transformation of the t-ZrO_2 to m-ZrO_2. This process was further elucidated by the work of Porter and Heuer[22] on a Mg-PSZ. To produce the optimum material, the precipitation of the t-ZrO_2 must be carefully controlled. This process involves the precipitation of oblate spheroidal, coherent precipitates on the {100} planes of the cubic phase and the aim of the heat treatment is to grow these precipitates to an optimum size. The increase in size changes the degree of metastability. If they are too large ($\gtrsim 0.1$ μm), the precipitates transform and this leads to microcracked materials. If they are too small, they are difficult to transform by stress and do not give very much toughening.

As pointed out earlier, these toughening concepts do not apply only to PSZ. There are now a wide range of transformation-toughened particulate composites based on ZrO_2, such as Al_2O_3-ZrO_2. Moreover, it was shown by Gupta et al.[23] that in systems such as CeO_2-ZrO_2 and Y_2O_3-ZrO_2, in which there is a fairly extensive solid solution region, single-phase tetragonal ZrO_2 materials can be produced and that these materials also possessed enhanced strength and toughness. A common feature of all these systems is that the ZrO_2 precipitate or grain size must be kept below a critical size in order to retain the tetragonal phase.

In summary, it was found that the tetragonal phase could be retained metastably at room temperature in many zirconia ceramics, provided the microstructure was carefully controlled. In addition, it was found that the phase transformation could be stress induced and that this led to a considerable increase in fracture toughness of the material. Indeed, fracture toughness values similar to cast iron are now feasible and remarkable strengths of up to 2.5 GPa have been reported. It is the understanding of these concepts and how they can be manipulated that is the primary aim of this book. Since these discoveries, the interest in zirconia has expanded considerably and many of the more recent developments have been discussed in special publications.[24-29]

In addition to the interesting refractory and mechanical properties previously mentioned, there are other properties which make ZrO_2 a particularly fascinating material. For example,

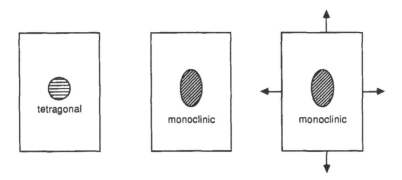

FIGURE 2. A schematic representation of the zirconia phase transformation. The normal phase transformation to monoclinic is represented by going from A to B and indicated that the zirconia particle undergoes a size and shape change. The material surrounding the particle will oppose the transformation and it is the strain energy that is involved in this constraint that allows the tetragonal phase to be retained. As shown in C, the transformation from A to C can be aided by an applied stress.

Nernst[30] used stabilized ZrO_2 as a glower for incandescent lighting and more recently ZrO_2 has been used as high-temperature heating elements, susceptors for induction heating, and electrodes for power generation by magnetohydrodynamics. These applications are a result of the ionic conductivity of ZrO_2. Wagner[31] established that the defect character of doped ZrO_2 was due to oxygen ion vacancies. Thus, the conductivity of these materials is the result of oxygen ion transport. The ability to conduct oxygen makes ZrO_2 important for many electrochemical applications. Major applications include oxygen sensors for control of automotive emissions, deoxidation of steel, combustion controls for furnaces and engines, electrochemical oxygen pumps, hydrogen production, and high-temperature fuel cells.

The optical properties of CSZ single crystals are also of interest since they possess a high refractive index. Indeed, one of the major uses of ZrO_2 has been as an opacifier for ceramic glazes. It is also possibe to grow large single crystals by skull melting.[32] These crystals appear very similar to diamond, especially as a result of the closeness in their refractive indexes. Thus, a flourishing industry for synthetic jewels, based on the production of cubic ZrO_2, has been established. The ability to grow large crystals of CSZ should lead to other potential uses. For example, tunable laser rods could be produced by appropriately doping CSZ.

IV. THE CONCEPT OF TRANSFORMATION TOUGHENING

It is worth reiterating some of the ideas discussed above to give a simple idea of transformation toughening. These ideas will be discussed in greater detail in the remainder of the book but these simple concepts should aid in developing a basis on which to expand.

A. Retention of Tetragonal Zirconia

Consider the metal oxide-ZrO_2 phase diagram shown in Figure 1. The diagram considers equilibrium conditions and, thus, one expects that the compositions that contain t-ZrO_2 at high temperatures will transform to m-ZrO_2 upon cooling. It was pointed out earlier, however, that if the t-ZrO_2 grain size is less than a critical size, it will remain tetragonal to much lower temperatures. The reason for this metastable retention is depicted schematically in Figure 2. Let us assume that the grain of t-ZrO_2 is at a temperature when it is expected to transform to m-ZrO_2. The phase transformation involves a set of transformation strains that increase the volume and change the shape of the particle. If the grain is isolated, the

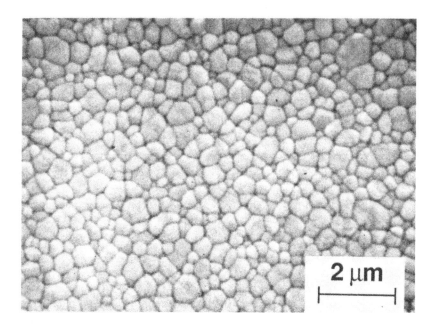

FIGURE 3.   Microstructure of a typical yttria (Y-) TZP material. The material is single phase t-ZrO$_2$ and the grain size is <1μm (thermal etch, scanning electron microscope, secondary electron mode). Micrograph courtesy of S. J. Glass, Pennsylvania State University.

transformation will occur provided the nucleation conditions are met. If, however, the grain is embedded in a matrix, there will be opposition to the transformation and there will be strain energy associated with the grain and the matrix around the grain. The production of this strain energy, which results from particle constraint, opposes the transformation by adding an extra term to the free energy of the system and makes the transformation more unlikely. In order to allow the transformation to proceed, the system has to be "supercooled" to a lower temperature, which increases the chemical driving force for the transformation. In some systems, the material can be cooled to room temperature and still the phase transformation has not occurred. The retention of t-ZrO$_2$ is a primary need for transformation toughening because it is this phase that leads to transformation toughening. The retention can be controlled by several microstructural and chemical factors, e.g., grain size and alloy content. An understanding of the crystallography of zirconia, its phase transformations, and how they can be controlled will be discussed in detail in Chapter 2.

B. Transformation-Toughened Microstructures

The earlier discussion mentioned several types of microstructures that can be used as a basis for ZrO$_2$-toughened ceramics (ZTC) and we can discuss these by considering the phase diagram in Figure 1. There are three different compositions marked A, B, and C shown on the diagram. Consider the thermal processing that occurs when these materials are first fabricated. The process usually involves forming a ceramic powder into the required shape and heating to a temperature at which densification occurs. For composition A, let us assume that it can be densified in the t-ZrO$_2$ phase field. The composition is then cooled to room temperature and, provided the grain size is less than the critical size required for the transformation, a **single-phase t-ZrO$_2$ material** will be produced. This microstructure is sometimes referred to as tetragonal ZrO$_2$ polycrystal or TZP and an example is shown in Figure 3. It is clear that only systems, in which there is a tetragonal solid solution, can be used to form TZP. Pure ZrO$_2$ or TZP can also be used as one of the components in a **composite**

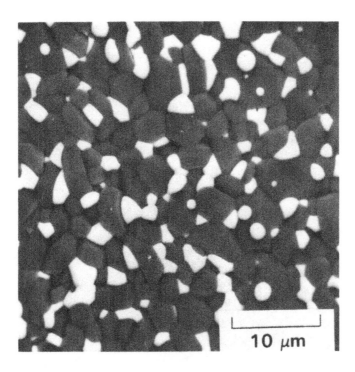

FIGURE 4. Microstructure of a zirconia (15 v/o) toughened alumina (ZTA) composite. The t-ZrO_2 grains exhibit the light contrast (thermal etch, scanning electron microscope, back scattering electron mode).

material and these are referred to as ZrO_2 dispersed ceramics (ZDC). An example of such a composite would be the Al_2O_3-ZrO_2 (zirconia-toughened alumina, ZTA) composites discussed earlier and a typical microstructure of ZTA is shown in Figure 4. In this approach, the composite systems are generally chosen so that the ZrO_2 does not react, or is chemically compatible, with the other phases. Route B shows a third possible approach, in which the ceramic is densified in the tetragonal + cubic phase field and again provided the t-ZrO_2 is less than the critical size, it will be retained to room temperature. The final microstructure consists of t-ZrO_2 and c-ZrO_2 as granular phases. Route C in Figure 1 shows the original approach that was used in the PSZ materials. The ceramic is densified in the single phase cubic region and then cooled down into the two phase tetragonal and cubic region. The ceramic is then given a specific treatment in this region to allow the t-ZrO_2 to precipitate and to grow to its optimum size. Thus, the final microstructure consists essentially of t-ZrO_2 precipitates in a cubic (c-) ZrO_2 matrix as shown in Figure 5. The different types of microstructure have been classified more carefully by Claussen[33] and this classification is shown in Figure 6. The different ZTC microstructures are schematically grouped into three classes. Group A ceramics (Row 1) are based on PSZ systems, i.e., those in which the ZrO_2 is alloyed with other oxides. Group B (Row 2) are those ceramics that contain ZrO_2 as a dispersed phase (ZDC), and Group C (Row 3) is a miscellaneous group, containing those microstructures that do not fit into Groups A or B.

C. Powder Processing

In the last section, the details of the fabrication procedure were glossed over, but as with most ceramics, it is usually the most critical part. As mentioned earlier, there are strict requirements for both the chemistry and microstructures of these materials, especially the

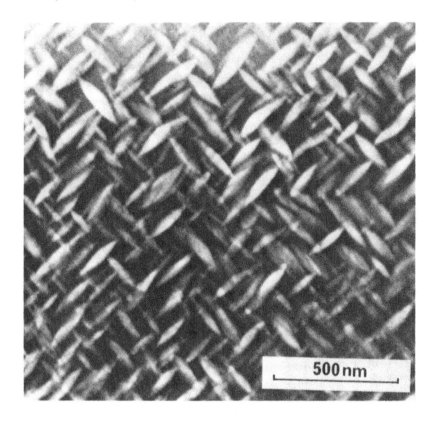

FIGURE 5.   Microstructure of a Mg-PSZ showing two orthogonal variants of the t-ZrO_2 precipitates (light contrast) within the cubic matrix. The third variant cannot be seen, but would be in the plane of the micrograph (ion beam thinned, transmission electron microscope).

final size of the t-ZrO_2 grains on precipitates. The ability to reach close to theoretical density, while maintaining a fine grain size, is very important in attaining maximum strength. Materials, such as Al_2O_3-ZrO_2 (Figure 4), in which one of the phases limits the grain growth of the other, represents one way to control grain growth. For other materials, recent innovations in the preparation of ultrafine powders, colloidal processing, and forming techniques have allowed the type of control that is needed. The techniques that are used to produce zirconia powders have been surveyed by Stevens.[6]

An important aspect of producing materials with maximum strength is to eliminate the various flaw populations that arise during the powder processing. For example, in ZTA, the major flaw populations have been found to be voids and agglomerates.[34] The voids may be due to incomplete sintering or the presence of organic impurities that were picked up during processing and then burned out during sintering. An example of the latter is shown in Figure 7A, in which the fracture origin was identified as a lenticular void, probably the result of the burnout of an organic impurity. Agglomerates, which may be present in the starting powder or may be due to poor mixing, lead to several effects during sintering. For example, the agglomerate may sinter differently than the surrounding material and lead to crack-like voids. Figure 7B shows an example in ZTA, in which a ZrO_2 agglomerate densified more rapidly than the surrounding material, leaving behind the void that caused failure. In some cases, the opposite may occur and the agglomerate does not sinter as quickly (Figure 7C). Another possibility is that the agglomerates may end up as large grains and, because of the low toughness of a single grain compared to a polycrystalline array, may lead to the formation of a crack (Figure 7D). It is worth pointing out that the identification of fracture origins is

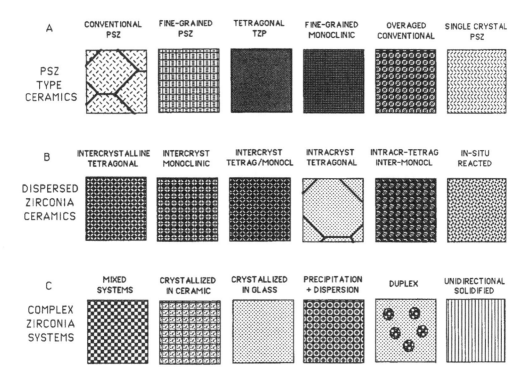

FIGURE 6. The classification of zirconia microstructures. (After Claussen, N., *Science and Technology of Zirconia II*, 1984, 325.)

an extremely useful tool in optimizing processing procedures. For some ZTC, however, the materials may not be as sensitive to the presence of processing flaws and the cause of this is discussed in the next section. In summary, it needs to be remembered that powder processing needs considerable attention in the production of the specific microstructures required in transformation-toughened ceramics.

D. The Source of Toughening

The final basic question is why does the stress-induced transformation lead to an increase in fracture toughness? If we return briefly to the t-ZrO_2 grain in Figure 2, we argued that the presence of a matrix around the grain opposed the transformation. If we were to stretch the matrix with applied stresses, it is clear that this would help the transformation to proceed and this is how the transformation can be stress induced. Figure 8 shows a crack passing through a transmission electron microscope specimen and the contrast is such that only m-ZrO_2 grains appear bright. It is clear that the stress field of the crack led to the transformation in a zone around the crack. If we now turn to Figure 9, a crack and its surrounding transformation zone are shown schematically. In some ways, the zone is like a large transformed inclusion that is constrained by the surrounding material. The transformation within the zone tries to make the zone larger, but this is opposed by the surrounding **untransformed** material. Thus, this latter material opposes the dilation of the transformed zone and presses back with residual stresses. The direction of these stresses are shown in Figure 9 and are acting to close the crack and hence reduce the driving force for crack extension.

The details of the mechanics involved in transformation toughening are more complex than discussed here and have been discussed in detail by several authors[22,35-39] and this will be the subject of Chapter 3. These studies led to some unusual aspects of crack growth behavior. It was found, for example, that transformation-toughened materials should show

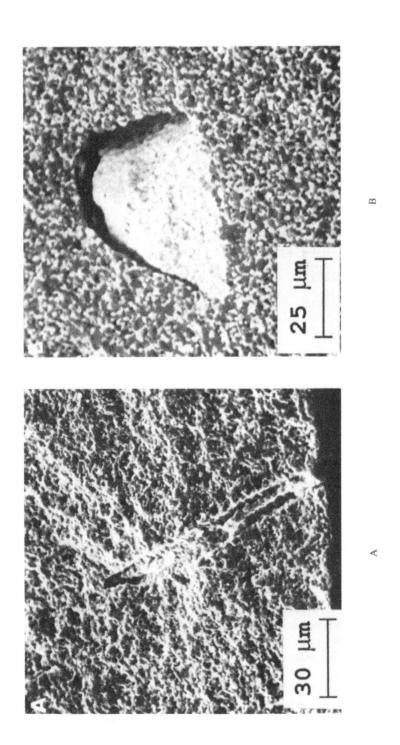

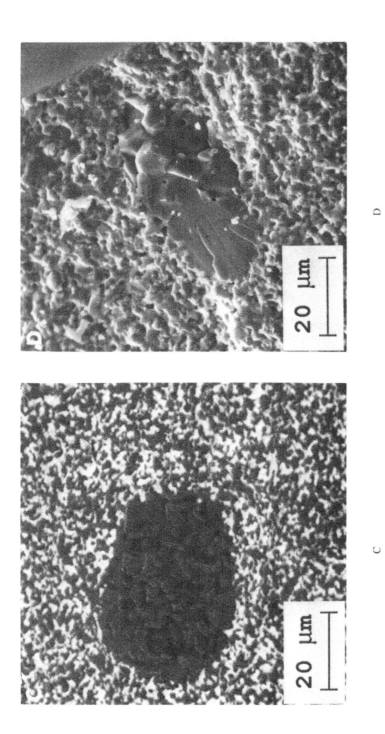

FIGURE 7. Scanning electron micrographs showing failure origins in ZTA: (A) lenticular void caused by burn out of an organic impurity (secondary electron image); (B) cracklike void caused by differential sintering of a ZrO_2 agglomerate (back-scattered image); (C) porous region associated with an Al_2O_3 agglomerate (back-scattered image); (D) large Al_2O_3 grain that originated from an agglomerate (secondary electron image). (Micrograph A courtesy of F. F. Lange, University of California, Santa Barbara.)

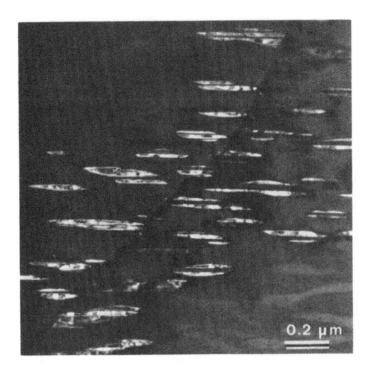

FIGURE 8. Transmission electron micrograph showing the region surrounding a crack that passes from the upper right to the lower left. The light contrast occurs only from zirconia precipitates that have transformed to monoclinic and these occur in a zone on either side of the crack. (Micrograph courtesy of L. H. Schonlein and A. H. Heuer, Case Western Reserve University.)

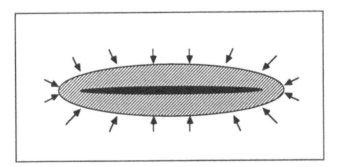

FIGURE 9. A schematic representation of the stresses that arise in a matrix that contains a zone (striped) around a crack (solid) when the zone undergoes a dilatant phase transformation. The residual stresses induced in the matrix around the zone "push back" and act to close the crack.

R curve behavior. That is, the resistance to crack growth should increase as the crack extends. This is a result of the size and shape of the transformation zone. In the initial stages of crack propagation, the stresses acting to close the crack are small; it is not until the transformation zone fully surrounds the crack that the maximum resistance to crack growth occurs. This discussion assumes that the transformation zone was formed irreversibly. Some recent work by Marshall and James[40] has shown that this is not always the case and there may be a reversible component to the transformation. The R curve effect is important in another way, in that it implies that small cracks can grow more easily than large cracks. Thus, some ZTC are insensitive to the presence of processing flaws. This flaw insensitivity is unusual for a ceramic material, but as pointed out recently by Swain and Rose[41] and Marshall,[42] it also leads to a limit in strengthening for a material of a given toughness. Although this limit may be seen as a barrier to the development of materials that are both strong and tough, it is expected to give flaw-tolerant materials with more uniform strength distributions, and this is particularly beneficial in the engineering design of these materials.

This discussion on the source of toughening has emphasized the transformation, but as pointed out earlier, microcracking has also been identified as an important toughening mechanism in ZrO_2 based materials.[19,43] Microcrack toughening can be thought of as a mechanism that involves the production of a large amount of fracture surface as a crack propagates, or because the microcracks lead to dilation in a zone around a crack. The dilation is a result of the opening of the microcracks and in many ways it is analogous to transformation toughening.[44] For example, if the transformation zone in Figure 9 was replaced with a microcrack zone, crack closure would again ensue, leading to a reduction of the crack extension force.[44] This crack tip ''shielding'' effect is also expected to give rise to R curve behavior as the process zone develops.[43,44]

There is one final aspect of the mechanical behavior of transformation-toughened ceramics that is worth mentioning. We have concentrated on the stress-induced transformation around a crack but it is also possible to induce the transformation on the outside surface of a specimen. This may be a useful attribute of these materials, in that the associated volume increase will place the outside surface in compression and could be used to strengthen the surface by making surface flaws less deterimental. This may be particularly useful in applications in which the surface has to withstand thermal and contact damage processes. It has been shown that the transformation can be induced by grinding,[45] impact,[46,47] and a variety of other techniques.[48]

The discussion in Section IV of this chapter can be thought of as the outline of this book. If one is to understand transformation-toughened materials, it is critical to understand the crystallography and phase relations in ZrO_2 (Chapter 2). Similarly, one must understand why the stress-induced phase transformation leads to increases in fracture toughness and how it is related to the microstructural parameters (Chapter 3). One must understand how to manipulate these parameters in the different types of transformation-toughened materials (Chapters 4 and 5). Finally, the nature of surface stresses, which are often present in these materials, must be understood (Chapter 6).

The remarkable increases in strength and fracture toughness that can be accomplished by transformation toughening has led to a new class of structural ceramics. This has opened the possibility of using these materials in many new applications, such as extrusion dies, cutting tools, bearings, and a wide variety of engine components, especially wear parts. In order to exploit the remarkable mechanical properties of these materials, however, it is critical for the engineers and scientists to understand the basic principles underlying transformation toughening.

REFERENCES

1. **Green, D. J.,** Industrial applications of ceramics, in *Industrial Materials Science and Engineering,* Murr, L. E., Ed., Marcel Dekker, New York, 1984, chap. 3.
2. **Gordon, J. E.,** *The Science of Strong Materials,* Penguin Books, Baltimore, 1973.
3. **Ryshkewitch, E.,** Zirconia, in *Oxide Ceramics,* 1st ed., Academic Press, New York, 1960, chap. II.5.

4. **Garvie, R. C.**, Zirconium dioxide and some of its binary systems, in *High Temperature Oxides*, Vol. 5-II, Alper, A. M., Ed., Academic Press, New York, 1970, chap. 4.

5. **Subbarao, E. C.**, Zirconia — an overview, in *Science and Technology of Zirconia*, Advances in Ceramics, Vol. 3, Heuer, A. H. and Hobbs, L. W., Eds., The American Ceramic Society, Columbus, Ohio, 1981, chap. 1.

6. **Stevens, R.**, Zirconia and Zirconia Ceramics, Magnesium Elektron Publ. No. 113, Magnesium Elektron, Twickenham, England, 1986.

7. **Ruff, O. and Ebert, F.**, Refractory ceramics. I. The forms of zirconium dioxide, *Z. Anorg. Allg. Chem.*, 180, 19, 1929.

8. **Ruff, O., Ebert, F., and Stephen, E.**, Contributions to the ceramics of highly refractory materials. II. System zirconia-lime, *Z. Anorg. Allg. Chem.*, 180, 215, 1929.

9. **Duwez, P., Odell, F., and Brown, F. H., Jr.**, Stabilization of zirconia with calcia and magnesia, *J. Am. Ceram. Soc.*, 35, 107, 1952.

10. **Stubican, V. S. and Hellmann, J. R.**, Phase equilibria in some zirconia systems, in *Science and Technology of Zirconia*, Advances in Ceramics, Vol. 3, Heuer, A. H. and Hobbs, L. W., Eds., The American Ceramic Society, Columbus, Ohio, 1981, chap. 2.

11. **Wolten, G. M.**, Diffusionless phase transformations in zirconia and hafnia, *J. Am. Ceram. Soc.*, 46, 418, 1963.

12. **Garvie, R. C., Hannink, R. H. J., and Pascoe, R. T.**, Ceramic Steel, *Nature (London)*, 258, 703, 1975.

13. **King, A. G. and Yavorsky, P. J.**, Stress relief mechanisms in magnesia and yttria-stabilized zirconia, *J. Am. Ceram. Soc.*, 51, 38, 1968.

14. **Garvie, R. C. and Nicholson, P. S.**, Structure and thermomechanical properties of partially stabilized zirconia in the CaO-ZrO_2 system, *J. Am. Ceram. Soc.*, 55, 152, 1972.

15. **Hasselman, D. P. H.**, Unified theory of thermal shock resistance of ceramic materials, *J. Am. Ceram. Soc.*, 52, 600, 1969.

16. **Green, D. J., Maki, D. R., and Nicholson, P. S.**, Microstructural development in partially stabilized ZrO_2 in the system CaO-ZrO_2, *J. Am. Ceram. Soc.*, 57, 136, 1974.

17. **Bansal, G. K. and Heuer, A. H.**, Martensitic phase transformation in zirconia (ZrO_2). I. Metallographic evidence, *Acta Metall.*, 20, 1281, 1972; II. Crystallographic aspects, *Acta Metall.*, 22, 409, 1974.

18. **Green, D. J., Maki, D. R., and Nicholson, P. S.**, in *Surfaces and Interfaces of Glass and Ceramics*, Materials Science Research, Vol. 1, Frechette, V. D., LaCourse, W. C., and Burdick, V. L., Eds., Plenum Press, New York, 1974, 369.

19. **Green, D. J., Embury, J. D., and Nicholson, P. S.**, Fracture toughness of a partially stabilized ZrO_2 in the system CaO-ZrO_2, *J. Am. Ceram. Soc.*, 56, 619, 1973.

20. **Bansal, G. K. and Heuer, A. H.**, Precipitation in partially stabilized zirconia, *J. Am. Ceram. Soc.*, 58, 235, 1975.

21. **Claussen, N.**, Fracture toughness of Al_2O_3 with an unstabilized ZrO_2 dispersed phase, *J. Am. Ceram. Soc.*, 59, 49, 1976.

22. **Porter, D. L. and Heuer, A. H.**, Mechanism of toughening partially stabilized zirconia (PSZ), *J. Am. Ceram. Soc.*, 60, 183, 1977.

23. **Gupta, T. K., Lange, F. F., and Bechtold, J. H.**, Effect of stress-induced phase transformation on the properties of polycrystalline zirconia containing metastable tetragonal phase, *J. Mater. Sci.*, 13, 1464, 1978.

24. **Heuer, A. H. and Hobbs, L. W., Eds.**, *Science and Technology of Zirconia*, Advances in Ceramics, Vol. 3, The American Ceramic Society, Columbus, Ohio, 1981.

25. **Claussen, N., Ruhle, M., and Heuer, A. H., Eds.**, *Science and Technology of Zirconia, II*, Advances in Ceramics, Vol. 12, The American Ceramic Society, Columbus, Ohio, 1984.

26. *Science and Technology of Zirconia, III*, Advances in Ceramics, The American Ceramic Society, Columbus, Ohio, in press.

27. *J. Am. Ceram. Soc.*, Special issue on zirconia, 69, March 1986.

28. *J. Am. Ceram. Soc.*, Special issue on zirconia, 69, July 1986.

29. **Evans, A. G. and Cannon, R. M.**, Toughening of brittle solids by martensitic transformations, *Acta Metall.*, 34, 761, 1986.

30. **Nernst, W.**, Electrolytic conduction in solid substances at high temperatures, *Z. Elektrochem.*, 6, 41, 1900.

31. **Wagner, C.**, Mechanism of electric conduction in Nernst glower, *Naturwissenschaften*, 31, 265, 1943.

32. **Aleksandrov, V. I., Osiko, V. V., Prokhorov, A. M., and Tatarintsev, V. M.**, in *Current Topics in Materials Science*, Vol. 1, Kaldis, E., Ed., North Holland Publishing, Amsterdam, 1978, 421.

33. **Claussen, N.**, Microstructural design of zirconia-toughened ceramics, in *Science and Technology of Zirconia, II*, Advances in Ceramics, Vol. 12, Claussen, N., Ruhle, M., and Heuer, A. H., Eds., The American Ceramic Society, Columbus, Ohio, 1984, 325.

34. **Lange, F. F.**, Structural ceramics: a question of fabrication, *J. Mater. Energy Syst.*, 6, 107, 1984.

35. **Evans, A. G. and Heuer, A. H.,** Review — transformation toughening in ceramics and martensitic transformations in crack-tip stress fields, *J. Am. Ceram. Soc.,* 63, 241, 1980.

36. **McMeeking, R. and Evans, A. G.,** Mechanics of transformation toughening in brittle materials, *J. Am. Ceram. Soc.,* 65, 242, 1982.

37. **Lange, F. F.,** Transformation toughening: part 1-5, *J. Mater. Sci.,* 17, 225, 1982.

38. **Marshall, D. B., Drory, M. D., and Evans, A. G.,** Transformation toughening in ceramics, in *Fracture Mechanics of Ceramics,* Vol. 5, Bradt, R. C., Evans, A. G., Lange, F. F., and Hasselman, D. P. H., Eds., Plenum Press, New York, 1983, 289.

39. **Budiansky, B., Hutchinson, J., and Lambroupolos J. C.,** Continuum theory of dilatant transformation toughening in ceramics, *Int. J. Solids Struct.,* 19, 337, 1983.

40. **Marshall, D. B. and James, M. R.,** Reversible stress-induced martensitic transformation in ZrO_2, *J. Am. Ceram. Soc.,* 69(3), 215, 1986.

41. **Swain, M. V. and Rose, L. R. F.,** Strength limitation of transformation-toughened zirconia alloys, *J. Am. Ceram. Soc.,* 69, 511, 1986.

42. **Marshall, D. B.,** Strength characteristics of transformation-toughened zirconia, *J. Am. Ceram. Soc.,* 69, 173, 1986.

43. **Green, D. J. Nicholson, P. S., and Embury, J. D.,** Microstructural development and fracture toughness of a calcia partially stabilized zirconia, in *Fracture Mechanics of Cermaics,* Vol. 2, Bradt, R. C., Hasselman, D. P. H., and Lange, F. F., Eds., Plenum Press, New York, 1973, 541.

44. **Evans, A. G. and Faber, K. T.,** On the crack growth resistance of microcracking brittle materials, *J. Am. Ceram. Soc.,* 67, 255, 1984.

45. **Pascoe, R. T. and Garvie, R. C.,** Surface strengthening of transformation toughened zirconia, in *Ceramic Microstructures, '76,* Fulrath, R. M. and Pask, J. A., Westview Press, Boulder, Colo., 1977, 774.

46. **Gupta, T. K.,** Strengthening by surface damage in metastable tetragonal zirconia, *J. Am. Ceram. Soc.,* 63, 117, 1980.

47. **Lange, F. F. and Evans, A. G.,** Erosive damage depth in ceramics: a study on metastable tetragonal zirconia, *J. Am. Ceram. Soc.,* 62, 62, 1979.

48. **Green, D. J., Lange, F. F., and James, M. R.,** Residual surface stresses in Al_2O_3-ZrO_2 Composites, in *Science and Technology of Zirconia, II,* Advances in Ceramics, Vol. 12, Claussen, N., Ruhle, M., and Heuer, A. H., Eds., The American Ceramic Society, Columbus, Ohio, 1984, 240.

Chapter 2

CRYSTALLOGRAPHY AND PHASE TRANSFORMATIONS IN ZIRCONIA AND ITS ALLOYS

I. INTRODUCTION

In this chapter we shall describe the crystal chemistry and the properties of zirconia which make it such an attractive material for use in transformation-toughening systems. Other properties which make it viable as an engineering ceramic will also be described.

The resurgent interest in zirconia is based primarily on its almost unique intermediate temperature transformation. This resurgence has been accompanied by a large number of publications which have contributed to an elucidation of the crystallographic nature of zirconia, its microstructural details, the influence of alloying additives, etc., all of which contribute to making zirconia such an academically interesting and appealing engineering ceramic. Notwithstanding this activity, there are still large areas of confusion, ambiguity, and lack of knowledge in our understanding of the behavior and characteristics of zirconia alloys and zirconia-containing systems. While it will not be possible to rexamine all the data and the premise upon which a large number of conclusions and hypotheses have been based, we shall attempt to present a coherent picture of the current state-of-the-art in zirconia crystallography and its general physical properties, phase transformations, alloy additives, and composite systems.

For a general background to the early work in zirconia, reference should be made to the excellent reviews by Ryshkewitch,[1] Garvie,[2] Subbarao et al.[3]

II. CRYSTAL CHEMISTRY

Zirconia (zirconium dioxide, ZrO_2) has long been used as a ceramic because of its refractory nature (mp 2680°C). Pure zirconia is known to exhibit three polymorphs, that is three crystal structures of the same or nearly similar chemical composition. Which structural form exists will depend upon temperature and pressure and to a lesser extent on stoichiometry. At atmospheric pressure, zirconia exists as a cubic structure above 2360°C; below this temperature, it is tetragonal. The tetragonal form is maintained to about 1200°C, below which the transformation to the monoclinic structure begins. Below about 1100°C, the structure is entirely monoclinic.

The polymorphs may be described by reference to the cubic fluorite type (CaF_2) structure, shown schematically in Figure 1. The tetragonal and monoclinic forms are considered distortions of the fluoride structure, and while the correlation is not strictly accurate, it nevertheless enables a ready comparison to be made of the three forms.

A. Monoclinic Zirconia

The first correct interpretation of natural (pure) zirconia (baddeleyite) was reported by Kathleen Yardley (Lonsdale) in 1926 as a monoclinic form with the space group $P2_1/c$.[4] Due to limited X-ray data, a definite crystal structure was not proposed. More than 30 years elapsed before McCullough and Trueblood[5] published the first accurate crystal structure data of monoclinic zirconia (baddeleyite). The R-value* of their unit cell was about 12%. A number of further refinements by Smith and Newkirk[6] reduced the R-value to about 9%. This R-value is poor by present day standards of crystal structure refinement. The reason

---

* R-value is the refinement value and is defined as the discrepancy between observed and calculated structure determinations.

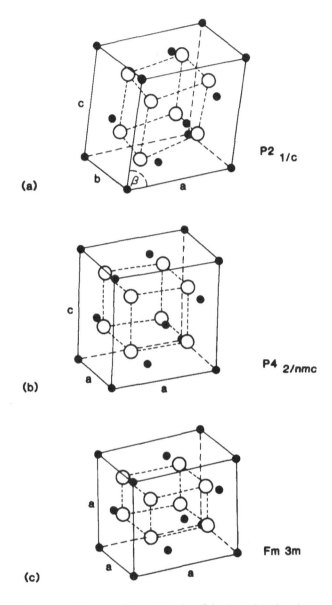

(a)

P2 1/c

(b)

P4 2/nmc

(c)

Fm 3m

FIGURE 1.   Schematic representation of the three zirconia polymorphs: a, monoclinic; b, tetragonal; and c, cubic. Their space
groups are indicated.

for this may possibly be due to the monoclinic zirconia forming a series of incommensurate
solid solutions[7] rather than a homogeneous single phase.[8] Alternate explanations are based
on the impurity of the natural baddeleyite, however, ultra-pure samples have yielded no
better structure refinements. The unit cell parameters of pure monoclinic zirconia are presented in Table 1 and an isometric projection based on the distorted calcium fluorite structure
is shown in Figure 1a. From the data in Table 1, it is evident that the lattice parameters are
yet to be refined sufficiently.

The monoclinic structure contains a number of interesting features. First, the Zr atom is
in sevenfold coordination with the O sublattice (compared to eightfold for the CaF_2 structure).
The idealized Zr coordination polyhedron in baddeleyite is shown in Figure 2. This figure
shows how the O coordination about Zr may be considered as being derived from a cube.

Table 1
LATTICE PARAMETER DATA OF THE
PURE ZrO₂ MONOCLINIC AND
TETRAGONAL POLYMORPHS

Monoclinic at room temperature

a	b	c	β	Ref.
0.5169 8	0.5232 8	0.5341 8	99°15'	5
0.5174	0.5226	0.5308	99°12'	9
0.5154	0.52075	0.53107	99°14'	104
0.51469 12	0.52056 11	0.53145 13	99°14'	105

Tetragonal at 1250°C

a	c	
0.364_{bct} [a]	0.527	10

Note: bct, body-centered tetragonal; fcc, face-centered cubic.
Values given in nm ± error in last digit(s).

[a] Equivalent to 0.515_{fcc}.

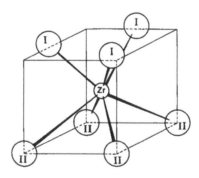

FIGURE 2. The "idealized" Z coordination polyhedron in baddeleyite. (After McCullough, J. D. and Trueblood, K. N., *Acta
Crystallogr.*, 12, 507, 1955.)

Similarly, the sevenfold coordination can be visualized by superimposing, on the Zr atoms,
in the {001} of the cube in Figure 1a, the O atoms of the planes above and below the Zr.

In Figure 2, the O_{II} atoms in the cube are tetrahedrally coordinated while the O_I atoms
on the other side of the cube are triangularly coordinated. Thus, the deviation from the cubic
fluorite structure is considerable, as illustrated by the distorted O configuration in Figure
1a. The O atoms vary in distance about Zr from O_I = 0.204 nm to O_{II} = 0.226 nm. A
second interesting feature is that the tetrahedrally bonded O_{II} layer has only one angle differing
significantly from the tetrahedral angle (109.5°). The variation is due to the triangularly
bonded O_I layer causing the ZrO₇ groups to be tipped over. This group configuration results
in a somewhat buckled nature of the O_{II} "plane", while the O_I plane remains planar within
experimental error. From models of the structure, it is also apparent that the Zr atoms in
the fluorite cell (f) faces are not only displaced from their $(^1/_2{}^1/_20)_f$ positions, but also that
the $(100)_f$ and $(010)_f$ are buckled due to the tetrahedral and triangular groups above and
below the Zr planes exerting a different ionic force on the atom. The Zr atom is displaced
out of the plane towards the tetrahedrally coordinated group.

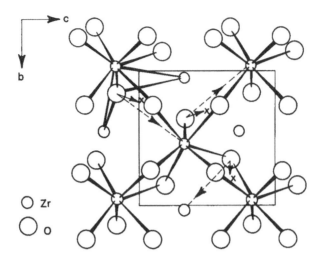

FIGURE 3. The layer of ZrO_2 groups at x = 1/4 projected on the $(100)_m$ plane. The unit cell outline represents its position at x = 0. The small x's show the oxygen positions in the tetragonal form. The small continuous arrows indicate the probable movement of oxygen according to Smith and Newkirk,[6] while the dotted arrow lines show the directions of probable atom movement according to Subbarao et al.[3] (After Smith, D. K. and Newkirk, H. W., *Acta Crystallogr.*, 18, 983, 1965.)

B. Tetragonal Zirconia

On heating, the monoclinic structure begins to transform to the tetragonal form at about 1200°C. The transformation is reversible and is accompanied by considerable hysteresis in temperature, normally occurring over a temperature range of about 100°C. It is the reverse transformation, tetragonal to monoclinic, which gives zirconia its desirable properties as an engineering ceramic. We shall examine the transformation more closely in the next section.

Ruff and Ebert[9] first observed the tetragonal form using a high-temperature X-ray camera. These workers firmly established the tetragonal form and showed it to be only a slight distortion of the analogous CaF_2 structure. Teufer[10] further refined the crystal structure and indexed it on the basis of a body-centered tetragonal (bct) lattice with space group $P4_2/nmc$. Teufer's lattice parameter determinations are shown in Table 1; the CaF_2 distortion of the tetragonal form is shown in Figure 1b.

The transformation of the monoclinic to the tetragonal form is not straightforward, (for the crystallography of the transformation see Section IV.D). The most significant event occurring during this transformation is the change in coordination of the Zr atoms from seven to eightfold. Figure 3 indicates the probable atomic route by which this increase in coordination is achieved.

Teufer's interpretation of the unit cell has two molecules in the elementary cell, comprising two sets of $Zr-O$ distances of 0.2455 nm and 0.2065 nm.[10] With the origin of the cell at $\bar{4}$ m2, the Zr and O atoms occupy positions similar to those in CaF_2 type structure; viz.

$$2Zr \text{ (a) } 0,0,0; \quad {}^1/_2,{}^1/_2,{}^1/_2$$

$$4O \text{ (d) } 0,{}^1/_2,z; \quad {}^1/_2,0,\bar{z}; \quad 0,{}^1/_2,{}^1/_2+z; \quad 0,{}^1/_2,\bar{z}$$

where z = 0.185. The R-value for the calculated structure factor is 6.7%. If the O atoms are placed in the exact CaF_2 configuration, i.e., z = 0.25, the R-value rises to 9.4% indicating an improbable configuration.

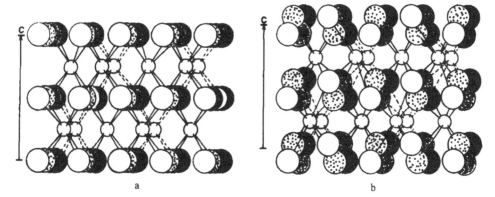

FIGURE 4. Tetragonal zirconia and its relation to fluorite: (a) tetragonally distorted fluorite and (b) the tetragonal structure of zirconia, (After Barker, W. W., Bailey, F. P., and Garrett, W., *J. Solid State Chem.*, 7, 448, 1973.)

The position of the O atoms produces a "rumpling" appearance of the distorted fluorite O (anion) layers. This situation is shown in Figure 4. Such a configuration of the O atoms gives rise to relatively weak "oxygen lattice" reflections in X-ray patterns. The strongest of these is the bct 102 reflection. This reflection can be used to uniquely identify the $P4_2/nmc$ space group of the "Teufer tetragonal" unit cell.[11] The intensity of these oxygen lattice reflections reduces rapidly as the number of anion vacancies increases. This may be achieved by additions of dopants or by raising the temperature. The additional effect of raising the anion vacancy concentration is to increase the value of z towards 0.25, so that the lattice may be considered to be approaching the cubic fluorite structure.

Since the original interpretation of Ruff and Ebert,[9] the lattice constants and indices have generally been described in terms of a face-centered tetragonal (fct) lattice. This annotation has persisted mainly because of convenience, particularly when comparing the tetragonal form with the face-centered cubic (fcc) CaF_2 high-temperature (or fully stabilized) structure in diffraction studies. In so doing, the tetragonal a-axis for the fct cell become $\sqrt{2}$ times the bct axis length. The unit cell indices are readily transposed using the standard body centered to face centered cell transform from the International Table of X-ray Crystallography, viz.:

$$\begin{pmatrix} h_t \\ k_t \\ l_t \end{pmatrix} = \begin{pmatrix} 1/2 & -1/2 & 0 \\ 1/2 & 1/2 & 0 \\ 0 & 0 & 1 \end{pmatrix} \begin{pmatrix} h_c \\ k_c \\ l_c \end{pmatrix}$$

C. Cubic Zirconia

Above about 2360°C, the tetragonal form of pure zirconia transforms to the cubic calcium fluorite structure (Fm3m), as shown in Figure 1c.[12,13] The transformation is diffusionless in that the lattice parameter changes by less than one atomic distance when the original tetragonal cell is considered in terms of the fct structure. As a result of the transformation, the z parameter in the unit cell becomes 0.25, so that all the O atoms take up the analogous CaF_2 configuration.

The existence of the pure stoichiometric cubic zirconia form, above 2360°C, is probably of academic interest, but has been questioned for sometime on a number of grounds. First, it was not possible to place the O atoms at z = 0.25 in the tetragonal cell and obtain a good R-value. Only the introduction of vacancies into the O lattice allows z to increase to 0.25. (This is the method employed to "stabilize" the cubic form to room temperature, see Section V.) Second, the ionic radius ratio of Zr to O is 0.57. At this value, the expected ionic cubic binary compound of the form MO_2 should have been of the rutile form, since the ionic radius ratio is in the range of 0.414 to 0.732.[14] As a result of these two discrepancies, factors other than simple ionic radius ratio must be considered when examining formation of cubic ZrO_2. One suggestion has been that significant covalent bonding may account for the CaF_2 form to be adopted.[15] Obviously the bond nature must play an important part in the atomic configuration and it is for this reason that vacancies also appear to play such an important role in the viability of a stable cubic-ZrO_2 structure. From the room temperature monoclinic form, it is evident that zirconia prefers sevenfold coordination to eightfold. It may be that the only way eightfold coordination can be accommodated is, if sufficient vacancies are present, to give a favorable charge balance.

III. PHASE TRANSFORMATIONS

An understanding of the mechanical properties of a material is generally described in terms of its microstructure. The microstructure of a material can include features ranging from the atomic scale up to the fabricated shape, i.e., perhaps from nanometer to meter scale. Since almost all useful engineering materials are composed of multiphase components, the most important parameter of a microstructure is possibly the "stability" of its various phases to changes by thermal, physical, or chemical influences. Microstructural instability has been appreciated by metallurgists for a long time. Two recent books by Martin and Doherty[16] and by Khachaturyan[17] give very good coverage of phase stability and transformations. We will present only a brief, rudimentary outline of the important facets of phase stability and transformations as they pertain to the materials under discussion and leave the reader to pursue the details in the referenced texts.

Transformation-toughened systems epitomize phase instability. In this sense, what we mean by phase instability is a thermodynamic instability such that the change in microstructure causes a decrease in the total free energy of the material and, thus, leads to a more stable structure. For most useful engineering materials, this change in free energy decreases the mechanical properties of a material, usually because the previous free-energy situation optimized the mechanical properties. However, in the case of transformation-toughened materials, the control of the stress-enhanced, thermodynamically driven transformation, to a *lower* free energy state, can significantly *increase* mechanical properties.

As described by Cahn for metallic systems,[18] there are only two possible types of instabilities which need be considered for systems undergoing transformations: one is a genuine instability, the other can be described as metastability. Cahn has presented a mechanical analogue of the two states. These are shown in Figure 5 where the unstable state is depicted by a wedge balanced on its tip and the metastable state situation consists of a rectangular block balanced on its small end. In the latter case, the transformation of the mestastable state to the final product, the system, must first overcome an intermediate, less-stable, and higher-energy state, before it can achieve a more-stable, lower state. In thermodynamic terms, the two situations are depicted by a transformation with and without a nucleation barrier. Atomistically, the two conditions may be considered as (1) the unstable state requiring the activation of a single atom while (2) the metastable state requires the activation of a number of atoms — a "nucleus".

In general, materials systems phase transformation occur by one of three routes: diffusional, interface controlled, and diffusionless or martensitic (see Section IV). The first two mechanisms are often termed civilian, while the third is referred to as military. For transformation-toughened systems, it is the martensitic reaction which imparts the benefit of enhanced mechanical properties. The civilian reactions occur, for example, in the decom-

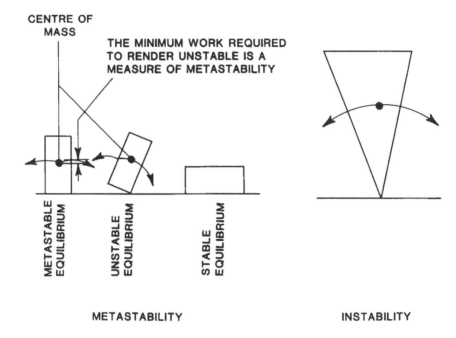

CENTRE OF MASS

THE MINIMUM WORK REQUIRED TO RENDER UNSTABLE IS A MEASURE OF METASTABILITY

METASTABLE EQUILIBRIUM

UNSTABLE EQUILIBRIUM

STABLE EQUILIBRIUM

METASTABILITY   INSTABILITY

FIGURE 5.   Representation of the difference between metastability and instability. (After Cahn, J. W., *Trans. Metall. Soc. AIME*, 242, 167, 1968.)

position reaction in magnesia partially stabilized zirconia (see Section V.B) and in the formation of ordered anion vacancy phases.*

A. Driving Force for the Transformation in Zirconia**

As will be apparent, the tetragonal to monoclinic transformation in zirconia can take place from a number of different states. For example, the tetragonal form may exist in the "free-state" as a single particle or crystal (see Section IV.D), or in the "confined-state" where it is constrained within a host matrix as either a precipitated or dispersed phase (see Section IV.E). Similarly, depending upon the size, stress situation, or location of the particle, its transformation may be easier or more difficult in the free-state than the confined state. The transformation has been addressed by a number of workers using different approaches such as nucleation or end-point thermodynamics, in an attempt to explain the critical factors controlling the initiation of the transformation. To date, the mechanism controlling the transformation is still not well understood, even those advocating nucleation mechanisms (see Section VI) cannot agree on a classical model (heterogeneous nuclei) or a nonclassical model (homogeneous nuclei, such as regions of high strain with diffuse interfaces) as the most likely nucleating source (see Section VI).

We can most easily visualize the transformation state of a material with the aid of free energy diagrams. The change in the free energy, i.e., the driving force for the transformation, can be represented as shown in Figure 6a, where the reaction coordinate is any variable defining the reaction path such as pressure. The free energy of the initial and final states are depicted as F_I and F_F respectively, such that the driving force $\Delta F = (F_F - F_I)$ and will

*   Diffusional transformation may occur at high temperatures in zirconia; although they may give rise to problems in long-term high-temperature applications, diffusional transformations are of little relevance to the low temperature transformation-toughening capabilities of zirconia-containing systems.[19,20]

** As pointed out by Martin and Doherty,[16] the term "force" is not strictly correct; however, in terms of the difference in the free energy associated with the transformed phases, we can consider the magnitude of this difference as the "driving force" for a reaction.

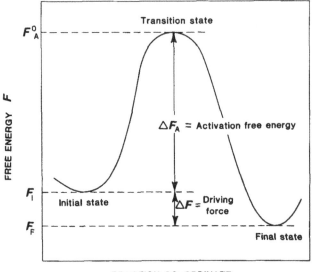

a

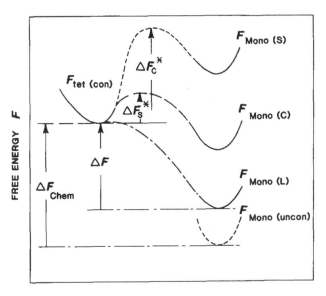

b

FIGURE 6.   (a) The change in free energy of an atom as it takes part in a transition. The reaction coordinate may be assumed as any variable defining the progress along the reaction path. (b) Schematic representation of the various free energy forms associated with the (Tet) tetragonal to (Mono) monoclinic transformation (normalized to the free energy of the initial tet phase), of (con) constrained Tet particles as a function of the initial particle size: S, small; C, critical; L, large, and uncon, unconstrained. See text for description.

be negative. The barrier which opposes the transformation is termed the activation energy ΔF_A. Until this energy is supplied, the system is said to be in a metastable state.

For zirconia this barrier can be overcome in two ways — thermally or stress (strain) assisted. It is the main aim of any fabrication process of transformation-toughened ceramics to develop the tetragonal zirconia phase into a metastable state, such that the nucleation barrier can be overcome by an applied stress, at or near room temeprature. The temperature at which the martensite transformation starts is known as M_s. Thus, the fabrication treatment must produce materials with an M_s temperature for the tetragonal phase just below room temperature so that spontaneous, i.e., thermally induced, transformation does not occur.

The ease with which the tetragonal to monoclinic transformation can occur in zirconia will depend upon a number of physical and chemical factors. Two of the most important factors are the particle size and the matrix in which the particles are being constrained (see Section V). The various energetic states of the tetragonal and monoclinic phases can be depicted schematically as shown in the free energy diagram of Figure 6b. On the left of the diagram is shown the free energy of a constrained tetragonal particle, $F_{Tet(con)}$, at room temperature T_O. On the right hand side of the figure is shown the free energies of the various states of the resultant monoclinic phase, F_{Mono}, as a function of the initial tetragonal particle size. At the top right of the diagram is the free energy of a small (confined) monoclinic particle, $F_{Mono(S)}$, at room temperature (i.e., its M_s is significantly below room temperature). In this situation, there is no net driving force for the transformation since the free energy of the constrained particle is higher than that of the tetragonal phase. Also in this situation, the activation barrier for nucleation of the transformation, ΔF_S^*, may be quite large and virtually insurmountable.*

A large particle, $F_{Mono\,(L)}$, whose M_s is above room temperature, will naturally transform upon cooling since ΔF_A, derived from thermal activation, will be exceeded when a particular temperature, sufficiently below M_s, is attained, i.e., sufficient undercooling is achieved. The intermediate case, depicted $F_{Mono\,(C)}$, for the transformation occurs when M_s for the particle is just below room temperature, i.e., the particle is a critical size. In this situation, ΔF_A may become very small and can be surmounted as a result of an applied stress.

It can now be appreciated that there exists a critical tetragonal particle size or state above which the particle will transform, either spontaneously or with the aid of an applied stress, and below which it will not. Therefore, as mentioned, it is the aim of processing and fabrication to provide an optimized volume fraction of "critically" metastable tetragonal particles to cooperate in the transformation-toughening process.

IV. MARTENSITE PHASE TRANSFORMATIONS

The phase changes from cubic to tetragonal and from tetragonal to monoclinic, in pure zirconia, may all be described as diffusionless or martensitic. The most-studied phase change in zirconia, because of its industrial significance, is the lower-temperature tetragonal to monoclinic transformation. Martensite transformations are of great industrial importance and have been studied in metallic systems for sometime.

Until recently the most important consequences of the martensite transformation in metallic systems was the development of a hard, strong metastable product, for example, by rapidly cooling Fe-C alloys. More recently its industrial application has been broadened to include the production of shape-memory alloys and now also ceramics. Many excellent reviews are available which describe the martensite transformation in metallic systems.[21-25]

---

* A situation can exist where an applied stress will induce the transformation even though the monoclinic phase has a higher free energy. Under these conditions the particle will often transform back to tetragonal upon removal of the stress, see Chapter 3, Section III.C.

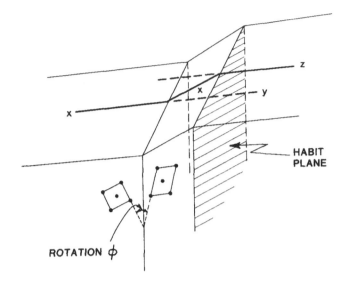

FIGURE 7.   Schematic representation of a martensitic invariant plane
strain transformation showing the surface tilting, scratch displacement, and
total lattice strain associated with the transformation.

As a prelude to the study of the martensite transformation in ceramics, especially zirconia
and its systems, we shall briefly summarize some of the properties which characterize such
transformations in metallic systems, predominantly Fe-C.

A. Characteristics of Martensite Transformations

Certain features characterize the martensite transformation. The transformation is diffusionless in that the movement of each atom involved in the transformation is less than one
atomic spacing. Being diffusionless, atomic displacement occurs by a cooperative (military)
movement of a large number of atoms by a shear process. Such a transformation results in
a composition invariant phase change, i.e., the initial and final (or parent and product) both
have the same composition. A further consequence of the diffusionless nature is that certain
crystallographic planes and directions within the parent will be common in the product, i.e.,
the transformation may be described by habit planes and directions. This feature is illustrated
in Figure 7. Being a diffusionless transformation, the velocity of the transformation may
approach that of sound within the crystal.

Shape change of a transforming crystal is often the most convenient experimental evidence
indicating that a martensitic reaction has taken place. Figure 7 illustrates the features of the
shape change associated with martensite transformations. A straight line or scratch present
on a plane surface, interesected by a martensite plate, remains straight and continuous, but
is changed in direction by a tilt. The surface remains plane but is tilted about its intersection
with the habit plane. To accommodate the rotation distortion between X and X′ (see Figure
7), both elastic and plastic deformation must be accommodated by the surrounding matrix.
Thus, the shape change may be described as homogenous lattice deformation, where the
three orthogonal directions (the principal strain axes) have remained orthogonal and remained
unrotated by the strain.[26] This homogenous lattice (strain) deformation implies a correspondence between lattice vectors in the parent and product of the transformation. Theoretically,
an infinite number of lattice correspondences may occur for a given transformation. In
practice, a correspondence is always observed and in theoretical treatments, it is assumed
that the lattice correspondence adopted will attempt to minimize the deformation strains.
When the formal crystallographic theories (see Section IV.B) are applied to any martensite

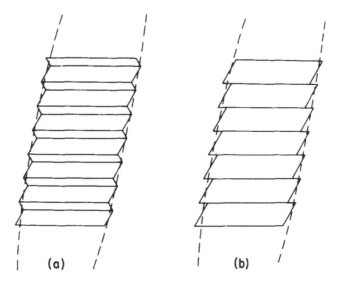

FIGURE 8. Schematic illustration of (a) internally twinned and (b) internally slipped martensitic plates. Such deformations ensure that the interface plane prevents the accumulation of strain and remains macroscopically undistorted. (After Wayman, C. M., *Introduction to the Crystallography of Martensitic Transformations*, MacMillan, New York, 1965.)

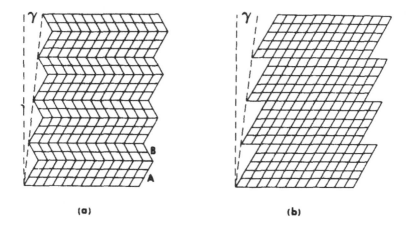

FIGURE 9. Schematic illustration of how slip and twinning may accomplish the same magnitude of inhomogeneous shear (angle γ).

transformation, an assumption of the lattice correspondence is an essential part in solving the crystallography. For many transformations, the lattice correspondence may be simply identified if the orientation relationship between the parent and product is known.

Experimental measurements have shown neither the habit plane, nor any line within it, is rotated by the shape strain (to within a few minutes of arc). This lack of rotation and macroscopic distortion of the habit plane is accompanied by a microscopic inhomogenous shear of the martensite plate. Figure 8 illustrates how this distortion may be accommodated within a martensite plate by either (a) internal twinning or (b) internal shear. Figure 9 shows how the same degree of inhomogenous shear may be accommodated by either (a) twinning or (b) shear.

The transformation is termed athermal and occurs over a marked temperature range. That is, once commenced, a martensite plate (or preexisting nuclei) will grow without the assistance of a thermally activated diffusion process and the amount of a martensite product resulting from the growth of a single nuclei is almost independent of time, but a function of temperature only. This temperature-dependent growth results because the strain set up by the shape and volume changes which accompany the growth reaction are not relieved by diffusion. The accumulating strain energy from the growth process opposes further transformation and so growth may stop before reaching completion. Further reduction in temperature (undercooling) below the M_s temperature provides an increased driving force to aid in the completion of the reaction.

The large strain energy associated with the martensite transformation explains the temperature range and hysteresis accompanying the forward (cooling) and reverse (heating) direction of the reaction. The martensite start and finishing temperatures in both the forward (M_s and M_f) and reverse directions (A_s and A_f), can be considerably influenced by composition and the stress (deformation) system acting upon the crystal.

The transformation is reversible, as mentioned above, in that the phases are the same before and after thermal cycling through the transformation.

B. Crystallography of Martensite Transformations

Two formal theories of martensite crystallography have developed independently from the original work of Greninger and Troiano,[27] namely, the Wechsler, Lieberman, and Read theory[28] and Bowles and Mackenzie theory.[29] The theories describe the phenomenological nature of the transformation using geometric and crystallographic aspects to account for the relationship between the parent and product. The theories state nothing about the motion of atoms which bring the change about. The two theories have used different mathematical procedures and approaches to solve the problem, but their results are the same.

In brief, the theories reduce the martensitic transformation of one phase into another using three physical steps:

1.   A lattice (Bain)[26] deformation
2.   A lattice invariant deformation (a simple shear)
3.   A rigid body rotation

Bain first recognized that a bct lattice could be transformed into a body-centered cubic (bcc) lattice by a small set of movements.[26] Figure 10 shows a bct cell delineated within a fcc lattice. The bct cell has an a-axial ratio of $\sqrt{2}$ times that of the fcc cell. A compression of the c-axis and an extension along the a-axis of the bct cell will result in the formation of a bcc lattice. This model was invoked to describe the austenite-martensite-ferrite reaction in Fe-C systems.

Any martensitic transformation such as the one just described in which pure homogeneous lattice deformation converts one lattice into another simply by expanding and contracting orthogonal axes is termed a Bain deformation. In practice, however, the operation cannot explain the oserved orientation relationships in the Fe-C system, and a rigid body rotation must also be involved in the axial ratio changes.

Thus, the rotation, when combined with the Bain transformation, is such that it leaves the habit plane *unrotated* and virtually *undistorted*. If this were not the case, misfit would occur between the transformation product and its matrix. This feature of the reaction then stipulates that a habit plane must be an invariant plane. When such a transformation occurs in a crystal, the associated strains are referred to as an invariant plane strain. While all three steps are required to affect the complete martensitic transformation, there is no specific time implied for the sequence or order of the transformation steps. With zirconia, since the

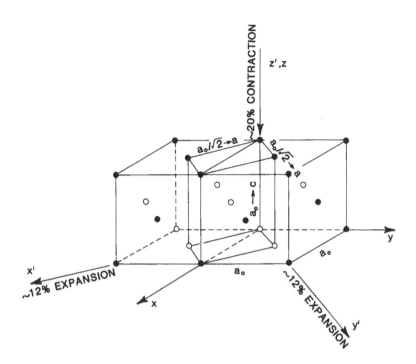

FIGURE 10. Lattice correspondence and distortion for the fcc to bcc (bct) transformation proposed by Bain.[26] The delineated bct unit cell is "upset" to the new dimensions during the martensitic reaction. The magnitudes of the principal distortions along the X', Y', and Z' are indicated. (After Wayman, C. M., *Introduction to the Crystallography of Martensitic Transformations*, MacMillan, New York, 1965.)

transformation often takes place in a host matrix, the associated strains cause highly localized stresses, which may result in matrix and/or interfacial cracking. These strains are another source of potential toughening (see Section VI.C and Chapter 3 Section IV.B)

If we take as an example the fcc → bcc (bct) transformation in Fe-C alloys (see Figure 10), it is possible to express the transformation via the Bowles-Mackenzie theory using matrix algebra.[29] The three main features of the transformation may then be expressed as matrix representations and may be written in the simplified form as:

$$P_1 = R \, B \, P \tag{1}$$

where P_1 is the shape deformation, R the rigid body rotation (ϕ in Figure 7), B the lattice (Bain) deformation, and P is a simple shear. The terms are expressed as (3 × 3) matrices. The rotation term (R) rotates the plane left undistorted by PB to its original position, thus Equation 1 represents an invariant plane strain.

The Bain distortion for the fcc → bcc, via bct, depicted in Figure 10 can be expressed as:

$$B = \begin{pmatrix} \dfrac{\sqrt{2}a}{a_o} & 0 & 0 \\[3mm] 0 & \dfrac{\sqrt{2}a}{a_o} & 0 \\[3mm] 0 & 0 & \dfrac{c}{a_o} \end{pmatrix}$$

Table 2
SHEAR TRANSFORMATIONS IN NONMETALLIC SUBSTANCES

Inorganic Compounds
  Alkali halides
    MX   (NaCl-cubic ⇌ CsCl-cubic)
  Ammonium halides
    NH_4X
  Nitrates
    $RbNO_3$ (NaCl-cubic ⇌ rhombohedral ⇌ CsCl-cubic)
    $TlNO_3$, $AgNO_3$, KNO_3 (Orthorhombic ⇌ rhombohedral)
  Sulfides
    MnS   (Zinc blende-type ⇌ NaCl-cubic)
    BaS   (NaCl-cubic ⇌ CsCl-cubic)
  Dicalcium silicate ($2CaO \cdot SiO_2$)   (Orthorhombic ⇌ monoclinic)

Minerals
  Pyroxene silicates
    Enstatite ($MgSiO_3$)   (Orthorhombic ⇌ monoclinic)
    Wollastonite ($CaSiO_3$)   (Monoclinic ⇌ triclinic)
    Ferrosilite ($FeSiO_3$)   (Orthorhombic ⇌ monoclinic)
    Olivine ⇌ spinel ($[Mg, Fe]_2SiO_4$)   (Orthorhombic ⇌ cubic)
    Quartz (SiO_2)   (Rhombohedral ⇌ hexagonal)

Ceramics
  Boron nitride   (Wurtzite-type ⇌ graphite-type)
  Carbon   (Wurtzite-type ⇌ graphite)
  Zirconia (ZrO_2), Hafnia (HfO_2)   (Tetragonal ⇌ monoclinic)

If typical values for three lattice parameters (a, a_o, and c) are substituted for the appropriate unit cells in the Fe-C alloys, the B matrix becomes:

$$B = \begin{pmatrix} 1.12 & 0 & 0 \\ 0 & 1.12 & 0 \\ 0 & 0 & 0.8 \end{pmatrix}$$

It can be seen that two of the principal distortions are greater than unity (x' and y' in Figure 10) and the third is less than unity (z' in Figure 10). Considering the "classical" Bain distortion, the necessary conditions for an invariant plane strain clearly do not exist.[24] A thorough introduction to the crystallography of martensitic transformations has been presented by Wayman[24] to which the reader is referred for an elaboration of the subject.

C. Martensite Transformations in Nonmetals

It is well established that some inorganic crystals, minerals, and ceramics achieve different atomic coordinations, bonding, or symmetry states by undergoing diffusionless transformations. Kriven[30] has made a comprehensive survey of shear transformations in such materials and has presented numerous examples, see Table 2. The table summarizes materials which are known to undergo large volume, coordination, or shape changes by a martensitic shear transformation mechanism. Some of these materials are possible candidates for fabrication into transformation-toughening systems.

In order to understand the details of martensitic transformation mechanisms involving large structure changes, three main features[31] of the transformation must be known:

1. Nucleation or initiation of the transformation conditions
2. The underlying structure correspondence or its absence
3. The structure of the interface after transformation and the mechanism of the misfit accommodation

As noted previously the analyses of martensitic transformations always involves features 2 and 3, feature 1 being almost impossible to observe or measure in most cases.

D. Transformations in Single-Crystal and Bulk Zirconia

Early work on the tetragonal to monoclinic transformation has almost invariably been performed either on pure "unconstrained" single crystals on or bulk material using transmission electron microscopy and high-temperature X-ray techniques. These studies showed the observed orientation relationships to be of the general form:

$$(100)_m//\{100\}_t \text{ and } [100]_m//\langle 100 \rangle_t$$

From this relationship, it is possible to identify three possible lattice correspondences (LC) depending upon which monoclinic axis is parallel to the tetragonal LC c_t-axis.[30] The possible correspondences are illustrated in Figure 11.

It has been difficult to distinguish between the possible lattice correspondences on the basis of experimental strain energy calculations. High-temperature transmission electron microscopy and X-ray studies[32-34] have confirmed LC B and LC C, accompanied by $(100)_m$, $(001)_m$, or $\{110\}_m$ twinning of the monoclinic product depending upon the transformation conditions (temperature, thin foil or bulk, etc.). The habit plane and lattice invariant deformation of the transformation have not been positively identified, although the following have been suggested by Bansal and Heuer for products of varying morphology.[33,34]

	Habit plane normals	Lattice invariant deformation
Lenticular-shaped product	$(671)_m$, $(761)_m$	$(1\bar{1}0) [001]$
Plate-shaped product	$(100)_m$	$(1\bar{1}0) [110]$

Full crystallographic analysis is complex and must be carried out with respect to the twin systems and lattice correspondences. A full analysis has not been reported to date. Information concerning the tetragonal to monoclinic transformation in a confining matrix is highly relevant to an understanding of the transformation-toughening effects of zirconia.

E. Transformation of Confined-Zirconia Particles

Tetragonal zirconia particles in a confined or constrained state may be of several forms:

1. Precipitates in partially stabilized zirconia (PSZ), e.g., Mg- and Ca-PSZ
2. Inter-and intragranular particles in zirconia-toughened ceramics (ZTC), e.g., zirconia-toughened alumina (ZTA)
3. Small grains, as in tetragonal zirconia polycrystals (TZP), e.g., Y- and Ce-TZP

The stability of the tetragonal forms, below M_s, will be determined by a number of factors as alluded to previously. We shall briefly examine the main ones here and consider them in more detail in Section VI. Some or all may be active in the three zirconia forms listed above. The major factors are

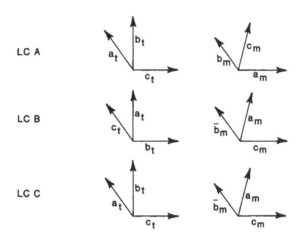

FIGURE 11.   Possible lattice correspondences for the zirconia tetragonal to monoclinic transformation. (After Kriven, W. M., *An International Conference on Solid-Solid Phase Transformations,* Aaronson, H. I., Laughlin, D. E., Sekerka, R. F., and Wayman, C. M., Eds., Carnegie-Mellon University, Pittsburgh, 1981, 1507.)

1.   Elastic moduli differences between the particle and matrix. This parameter will give rise to a term in the free energy equilibria and is reflected in the shape and volume change which must be accommodated by the matrix or particle at the time of transformation.
2.   Interfacial energies due to coherent, semi-, or incoherent particle matrix interface and thermal expansion coefficient mismatch may influence the nucleation and driving force of the transformation.
3.   Particle size, morphology (shape), and location will influence M_s, transformation nucleation mechanisms (e.g., thermally or stress induced), and the transmission of stress within the particle will also play a part.
4.   An applied stress may cause very significant variations in M_s, by overcoming the nucleation barrier, ΔF_A.

Unlike bulk or large-grained zirconia materials, the three forms of constrained zirconia particles may generally be prepared in such a way (as we have seen in Section III) that the transformation can be studied at low temperatures, i.e., M_s temperatures are at or below room temperature. Then, by using careful transmission electron-optical techniques, the crystallography of the parent and transformation product may be studied. Such studies provide an experimental determination of parameters for use in strain calculations. Before we examine specific cases some mention should be made of the transformation nucleation mechanisms.

As with metallic systems, the actual initiation or nucleation mechanisms for the transformation in zirconia is not well understood. For small free or confined particles, several nucleation mechanisms have been proposed. All models must satisfy the basic criterion whereby a sufficiently large monoclinic nucleus is created for which the monoclinic phase may grow (or continue to propagate). This nucleation must satisfy thermodynamically favorable conditions i.e., the energy required to form the nucleus must not be so high that it is physically unrealistic. In most models, the energy requirements to create a critically sized nucleus is too high. However, two models warrant consideration. Both are based on earlier models proposed for martensitic transformations in metal systems. The models have been adapted for zirconia by Chen and Chiao[35] and by Heuer and Ruhle.[36]*

Chen and Chiao[35] studied the statistical probability of martensite nucleation in zirconia-containing ceramics and in Fe-Ni alloys as a function of particle size and temperature. Their model depends upon the classical heterogeneous nucleation process which develops from an inherent defect, leading to a critical nucleus size. The critical nucleus may occur from an intrinsic, preexisting defect, or an extrinsically stress-induced nucleus. Which nucleation type causes the transformation (both types must be stress assisted) is related to a size effect via a "potency distribution." The potency distribution is related to the probability of finding a defect of unit height, which extends indefinitely in one direction (e.g., a dislocation). This defect contains along its length a shear strain of the order of the martensitic transformation strain. The statistics, then, are related to the probability of finding such a critically sized defect. While the authors produce experimental evidence, by observing induced defects in transmission electron microscope experiments, general observations on fabricated materials have failed to reveal obvious nucleating defects. For example, no such nucleating defects have been observed when an apparently featureless zirconia particle is transformed by beam heating in a transmission electron microscope.

The inability to observe obvious nucleating defects, as demanded by the classical nucleation theory, has caused Heuer and Ruhle[36] to propose another model based on nonclassical nucleation. These workers are not persuaded by the observations and arguments of Chen and Chiao,[35] but prefer, instead, to consider the nucleation process in terms of a nonclassical localized soft mode (LSM) model proposed by Guenin and Gobin[38] for metals, which was originally based on a model proposed by Clapp.[39] The LSM model requires lattice vibrations (or phonons) to introduce critical lattice strains (strain spinodals) along directions of low elastic modulus (soft directions) within a crystal, so that the lattice may be distorted to the new phase structure and the only barrier to critical martensite nucleus formation is the interfacial energy (between parent and product). The model relies on a diffuse interface and critical strains being reached at certain types of defects. The defects thought most favorable include free surfaces, interphase interfaces, strain sites from coherency stresses, thermal expansion aniostropy, dislocations, stacking faults, and, to a lesser extent, point defects. As will be appreciated, fabricated materials contain all these defects to a greater or lesser degree, so that a nucleus in the Clapp sense will always be available for activation by an applied stress. The model,** while used to explain the nucleation phenomenon, has a major drawback in that no second or third order elastic constants are available for zirconia, so that it has not been possible to calculate the possible soft mode directions and magnitudes.

All models are in agreement that nucleation is always stress assisted. The stresses may arise from inherent causes, such as lattice mismatch or thermal expansion anisotropy, or from externally applied sources (stress concentrations).

F. Transformation of Confined Particles in Partially Stabilized Zirconia (PSZ)

Of the precipitated zirconia systems described in Section V, the Mg-PSZ system is the most suitable for studying the crystallography of the tetragonal to monoclinic transformation in a confining matrix. In this material, the tetragonal precipitates are lenticular in shape (see Figure 12a and discussion in Section IV) with the c_t-axis always parallel to the axis of rotation of the precipitate plate, while the plate habit plane is always $\{100\}_c$.

---

* Anderson and Gupta[19] have considered a nucleation process for the transformation based on preexisting "embryos". The embryo size they suggest is quite large (3.5×10^7 unit cell; Heuer et al.[37]), which would be detectable by transmission electron microscopy. Unfortunately, no such nucleation sites have been observed, so their model must be considered to be only tentative.

** Garvie and Burke[40] also invoked the Clapp model to explain the nucleation of the martensitic tetragonal to monoclinic transformation in ZrO_2 and HfO_2.

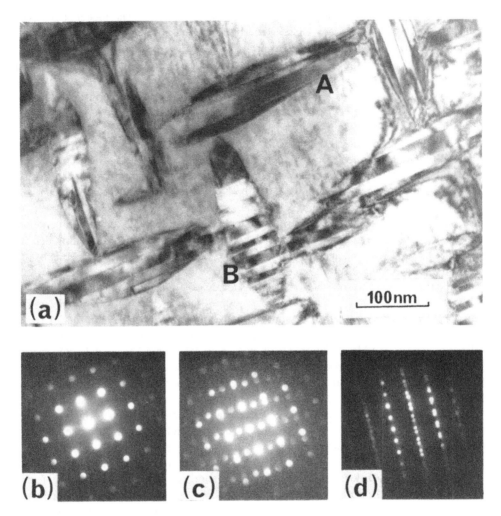

FIGURE 12. (a) Transmission electron micrograph (bright field) showing dispersed m-ZrO₂ particles in an "over-aged" specimen of 9.5 mol % MgO-ZrO₂. (After Muddle, B. C. and Hannink, R. H. J., *J. Am. Ceram. Soc.*, 69, 547, 1986.) (b) Selected area electron diffraction pattern of cubic matrix phase in (a), beam direction ⟨100⟩c. (c) Microdiffraction pattern (MDP) from adjacent pair of monoclinic variants in particle B. A schematic solution is provided in Figure 14. (d) MDP from a single monoclinic variant in particle A. The schematic solution is provided in Figure 13.

Using convergent beam microdiffraction techniques, Muddle and Hannink[41] have presented the first detailed report of the tetragonal to monoclinic transformation crystallography of athermally and stress-assisted transformed tetragonal precipitate particles in Mg-PSZ.* We shall summarize their observations.

The typical athermal or stress-assisted monoclinic transformation product has a substructure comprising parallel variants of the monoclinic phase, as demonstrated in Figure 12. As seen in this figure, the monoclinic variants extend either parallel (particle A) or normal (particle B) to the habit plane of the original tetragonal particle and the adjacent pairs are twin related — to within the accuracy of the electron microdiffraction technique. Figure 13

---

* An orthorhombic phase, occurring as an intermediate phase between tetragonal and monoclinic phase transformation, has been identified in Ca- and Mg-PSZ.[42-44] However, it has been demonstrated[45] that this phase occurs as a thin foil artifact and has not been demonstrated to take part in bulk transformations.

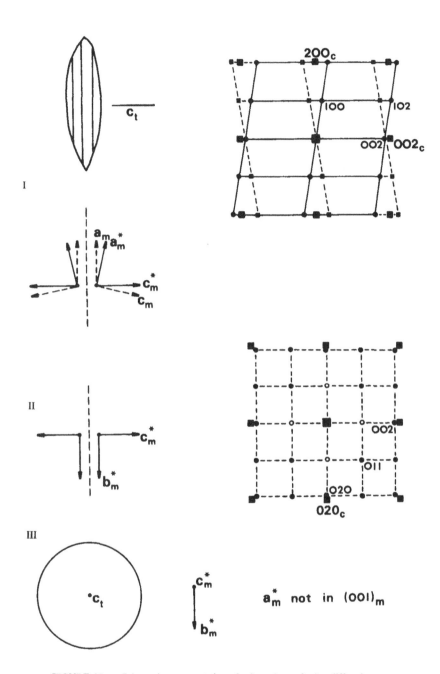

FIGURE 13a. Schematic representation of orientations of microdiffraction patterns for (I) particle A in Figure 12; (II) equivalent monoclinic particles such as particle A but where the particle is rotated 90° with respect to I; (III) a monoclinic particle such as particle A situated normal to the electron beam direction. Open circles in patterns indicate forbidden reflections that may occur as a result of double diffraction. (After Muddle, B. C. and Hannink, R. H. J., *J. Am. Ceram. Soc.*, 69, 547, 1986.)

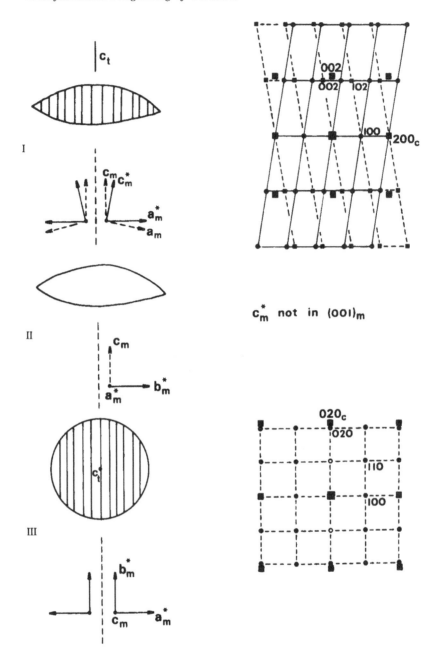

FIGURE 13b.   Schematic representation of monoclinic orientations and micro-diffraction patterns (I) obtained from particle B in Figure 12; (II) equivalent monoclinic particles rotated 90° about c_m with respect to that shown for I; and (III) particle normal to the electron beam. Open circles in patterns indicate forbidden reflections that may occur as a result of double diffraction. (After Muddle, B. C. and Hannink, R. H. J., *J. Am. Ceram. Soc.*, 69, 547, 1986.)

shows the schematic representation of the monoclinic orientations and corresponding indexed microdiffraction patterns. From these figures, it is evident that for particles with twin variants in $(001)_m$, particle A, the orientation relationship between the tetragonal and monoclinic lattices is such that:

$$(001)_m//(001)_t \quad \text{and} \quad [100]_m//[100]_t$$

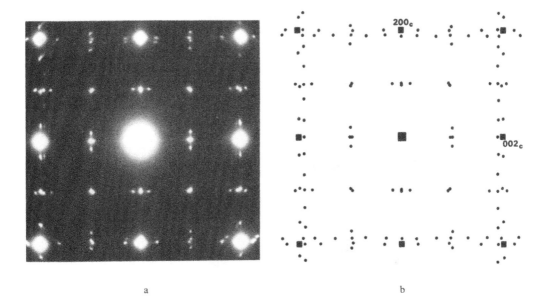

a

b

FIGURE 14. Comparison of (a) observed $\langle 100 \rangle_c$, selected area electron diffraction pattern with (b) schematic pattern formed by superimposing the single-crystal patterns expected of 12 permitted monoclinic orientations and their true variants. (After Muddle, B. C. and Hannink, R. H. J., *J. Am. Ceram. Soc.*, 69, 547, 1986.)

In precipitates with the transverse twin variants (Figure 13b), the boundaries are parallel to $(100)_m$, particle B, and the orientation relationship is given by:

$$(100)_m//(100)_t \quad \text{and} \quad [001]_m//[001]_t$$

While these relationships differ by a rotation of approximately 9° about $[001]_t$ in each case, the lattice correspondence is such that c_m is parallel to c_t. Therefore, the transformation prefers to adopt the LC C configuration shown in Figure 11 and the correspondence matrix can thus be expressed in the following equation as originally proposed by Bansal and Heuer.[33,34]

$$_mC_t = \begin{pmatrix} 1 & 0 & 0 \\ 0 & 1 & 0 \\ 0 & 0 & 1 \end{pmatrix}$$

A total of 12 possible monoclinic orientations (plus the two variants of each) can be detected in a given orientation of the cubic phase. These orientations are derived from the original three variants of the tetragonal, two possible orientation relationships between tetragonal and monoclinic lattices, and two possible variants in each relationship. A comparison of the observed $\langle 100 \rangle_c$ electron diffraction pattern, obtained from an area shown in Figure 12a, and a schematic of a pattern formed by superimposing all 12 expected monoclinic orientations and their twin variants, is shown in Figure 14.

The possible monoclinic orientations apply both to particles transformed athermally and to those transformed under stress. The only criterion for preferred monoclinic orientation is that, in any given volume of material and fraction of particles adopting a particular orientation of monoclinic structure, it is strongly influenced by the conditions under which the transformation occurs. Transformation cycling experiments carried out in a transmission electron

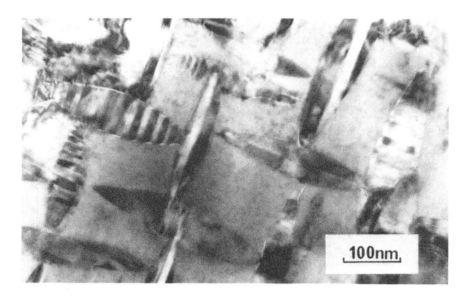

FIGURE 15.   Bright field transmission electron micrograph illustrating microcracking associated with transformed ZrO$_2$ particles. (After Muddle, B. C. and Hannink, R. H. J., *J. Am. Ceram. Soc.*, 69, 547, 1986.)

microscope lead to similar conclusions, in that the twin/variant system configuration adopted by the monoclinic phase is one where the residual transformation strains are minimized.[46]

Kelly and Ball[47] have used the phenomenological martensite theory of Bowles and Mackenzie[29] to calculate the possible preferred crystallographic relationships of the tetragonal to monoclinic transformation in the Mg-PSZ system just described. Their calculated predictions are in very good agreement with the observations just described.

Microcracks, which also contribute to the overall toughening in ZTC (see Chapter 3 Section IV.B), are observed in the Mg-PSZ materials discussed here. The preferred sites for microcrack nucleation occur at the intersection of the monoclinic variant boundaries with the precipitate-matrix interface (see Figure 15). How this situation arises is schematically depicted in Figure 16. The microcracks are assumed to be a product of the transformation (rather than a foil-preparation artifact) and are found to accommodate lattice strains arising at the precipitate boundaries. In bulk material, where thermal transformation has occurred, a toughening increment is still possible (see Chapter 3) if the advancing crack induces microcrack propagation from the highly strained monoclinic variant-precipitate-matrix triple-point sites.

A similar analysis for transforming tetragonal precipitates in Ca-PSZ and Y-PSZ is not so readily accomplished since the particle morphology does not lend itself to a ready interpretation of the cubic, tetragonal, and monoclinic relationships. However sufficient evidence is available to suggest that the transformation and associated twin modes will also be dependent upon the stress system operating at the time of transformation.[48]

G. Transformation of Confined Particles in ZTC Systems

The dispersed zirconia phase in ZTC generally consists of irregular and regular shaped zirconia particles dispersed in inter- and intragranular positions, respectively. While α-Al$_2$O$_3$ is the most common matrix used, β-Al$_2$O$_3$, mullite (3Al$_2$O$_3$·2SiO$_2$), silicon nitride, silicon carbide, spinel (MgO·Al$_2$O$_3$), and other more experimental systems (ThO$_2$, ZnO, MgO, TiB$_2$, titanium carbonitride) have also been successfully toughened by the incorporation of particulate zirconia, see Chapter 5.

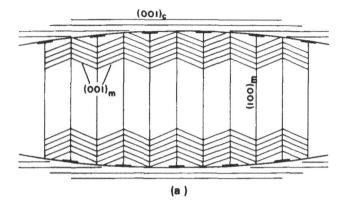

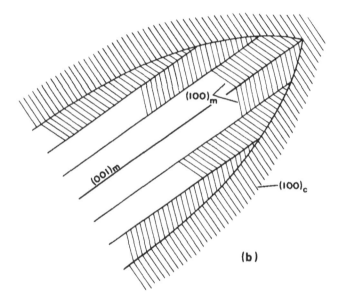

FIGURE 16. Schematic representation of lattice displacements developed at the particle matrix interface as a result of the tetragonal to monoclinic transformation. Particles have twin variant boundaries parallel to (a) $(100)_m$ planes and (b) $(001)_m$ planes. Positions of maximum strain are potential sites for microcrack nucleation. (After Muddle, B. C. and Hannink, R. H. J., *J. Am. Ceram. Soc.*, 69, 547, 1986.)

The transformation crystallography of dispersed zirconia particles is difficult to study using transmission electron microscopy because thin foil effects usually do not allow the tetragonal form of intergranular particles to be retained, while intragranular particles are also susceptible to transformation in thin foils. The most detailed study of the transformation crystallography, of intragranular zirconia particles has been presented by Kriven.[49] Using a very specific case and experimental observations of many samples, Kriven was able to show first, that small ($\sim$0.3μm) tetragonal zirconia particles constrained within a α-Al$_2$O$_3$ matrix were in a state of tension, due to the thermal expansion anisotropy.* A misfit strain of $\sim$1.2 $\times$ 10^{-2} was calculated to exist between the two phases. Prior to the transformation, the

* Typically for specimens fabricated at 1500°C with thermal expansion coefficient of Al$_2$O$_3$, α = 8.1 $\times$ 10^{-6} °C^{-1} and for t-ZrO$_2$ α_a = 11.6 $\times$ 10^6 °C^{-1} and α_c = 16.8 $\times$ 10^{-6} °C^{-1}.

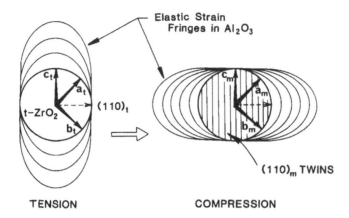

FIGURE 17.   Suggested transformation and deformation twinning mechanism for a transforming ZrO_2 particle in an α-Al_2O_3 matrix. The model, proposed by Kriven,[49] is consistent with a principal tetragonal strain axis lying parallel to $[001]_t$ (a) and the occurrence of LC C, and the principal monoclinic strain was perpendicular to $(110)_m$ twins (b). (After Kriven, W. M., *Science and Technology of Zirconia II*, 1984, 64.)

principal axis of the strain field is in the c_t-direction of the particle, as shown schematically in Figure 17. Second, when such a particle is transformed in the foil, as shown schematically in Figure 17, it does so according to the LC C mechanism, i.e., lattice correspondence a_t, b_t, and c_t to a_m, b_m, and c_m is preserved and twinning occurs on $(110)_m$. The principal monoclinic strain field is then perpendicular to the $(110)_m$ twin plane for these small particles.* This mode of transformation also suggests for ZTC systems, as with Mg-PSZ, the transforming particle is pursuing a route which will minimize its shape change during the transformation.

Crystallographic analyses on these systems is by no means complete, for when Kriven attempted to calculate the preferred transformation and twinning modes according to the generalized CRAB (Crocker-Ross-Acton-Bevis) theories,[50,51] it was found that the theory did not support the experimental observations. This observation lead Kriven to suggest that either more careful physical measurements are required to test the equations, or that the phenomenological theories of martensite transformation may need modification when applied to transformations involving confined particles.

The transformation twin modes of ZrO_2 confined in a mullite matrix has been presented by Bischoff and Ruhle.[52] These workers studied the twin configuration of a large, thermally transformed ZrO_2 particle and found that (100), (110), (001), and (011) twins were all present. Very often (100) and (110) twins formed a mosaic structure within a particle. The occurence of (110) twins often occurred as "closure" twins at the end of larger (100) twins. These closure twins help to minimize the shape change and the strain at the particle-matrix interface. If they are not present, as found for the Mg-PSZ material, microcracks may form, depending on the particle size (see Figure 15). Thus, Bischoff and Ruhle[52] concluded that the complicated twin structures, observed in thermally transformed zirconia particles constrained in a mullite matrix, all resulted from an accommodation process which minimizes the strain associated with the transformation.

---

* Larger particles, transformed during cooling to room temperature, showed twinning on (100) while another was twinned on (001). These observations again indicate that, depending on the transformation environment, all previously reported twin systems in ZrO_2 can occur during the transformation of confined particles.

H. Transformation of TZP Materials

A complete crytallographic analysis of the tetragonal to monoclinic transformation of TZP materials has not yet been reported. The main studies involving the transformation have described the nucleation and propagation of the monoclinic phase in tetragonal grains.[53,54] It is anticipated, however, that the twin systems would be similar to those observed for other systems where the zirconia particles are confined within a host matrix. The main difference in TZP systems is that the body is composed almost entirely of metastable tetragonal grains. In this situation, when one grain transforms, the accompanying transformation strains are immediately experienced by the adjacent metastable grain, and a situation of "autocatalytic" nucleation of the transformation in the adjacent grain may then follow. This phenomenon has been observed (see, for example, Ruhle et al.[53]).

More recently, evidence has been obtained in a 12 mol % CeO_2-ZrO_2 TZP alloy that the transformation, occurring around a propagating crack, does not transform the whole grain to the monoclinic form.[55] It appears, therefore, that the deformation around the crack is accommodated by regions (sheets) of monoclinic phase interspersed with regions (blocks) of tetragonal. Hence the toughness is more closely related to a situation of "transformation plasticity" rather than transformation toughening, in that a very small volume fraction of the grain is actually transformed to monoclinic, yet considerable plastic deformation can be accomodated (see also Chapter 4 Section IX).

V. ALLOY ADDITIVES FOR ZrO_2

Pure zirconia could not be used for fabricated ceramic forms. Therefore, it has been the custom for at least 50 years to add "stabilizers" to zirconia to retain the high-temperature cubic-fluorite-type phase. Stabilizers found most suitable are aliovalent cubic oxides with cation size ratio differences of <40% that of zirconia, when the coordination number in the lattice is eight, i.e., zirconia in the cubic or tetragonal phase. Table 3 presents some of the common alloying elements and their percentage of ionic radii difference with respect to zirconia.*

The actual stabilization mechanism is not well understood. In simple terms, stabilization may be brought about by an oxide with a room-temperature-stable cubic structure, causing the distortion of the tetragonal and monoclinic phases back to the cubic form, as suggested by Ryshkewitch.[1] Alternatively, the removal of anions providing a suitable electronic charge balance will allow the cubic structure to survive, as mentioned in Section I. Along similar lines, Barker and Williams[56] have suggested that the nearer the ionic radius of an altervalent stabilization cation is to that of Zr^{4+}, the more effective its stabilizing "efficiency" in terms of minimum mole per cent required to yield full stabilization. Thus, Sc^{3+} is suggested as the most effective trivalent stabilizer because of the close ionic radii (see Table 3). The reason for its stabilization effectiveness, other authors suggest, is due to the ease of anion ordering while the cation lattice remains disordered, such that the zirconia lattice may adopt a more stable state. The inference here is probably the eightfold coordination of the Zr^{4+} which will be adopted on stabilization to tetragonal or cubic symmetry, thereby bringing it closer to a true "ionic" solid.

Traditionally magnesia (MgO), calcia (CaO), and yttria (Y_2O_3) have been used as stabilization oxides in commercial materials. Other oxides such as ceria (CeO_2), scandia (Sc_2O_3), lanthinum oxide (La_2O_3), and ytterbium oxide (Yb_2O_3) are also being used successfully.

By using suitable stabilizer additions, two forms of transformation-toughening zirconia may be produced. These systems may be classified as:

---

* The "physical" size that a cation adopts withing a host lattice will naturally depend upon the valency it adopts, i.e., the spread of the electron cloud. Therefore, the ionic size shown in Table 3 is merely a rough estimate, as a first approximation, the size ratio correlates very well with this "stabilization" size effect.

Table 3
IONIC RATIO OF ELEMENTS
OFTEN ALLOYED WITH
ZIRCONIA, SHOWN FOR A
COORDINATION NUMBER OF 8

Element	Ionic radius (Angstroms)	Difference WRT ZrO_2
Zr^{4+}	0.84	—
Ba^{2+}	1.42	+69%
Ca^{2+}	1.12	+33%
Ce^{4+}	0.97	+15%
Hf^{4+}	0.83	−1%
Mg^{2+}	0.89	+6%
Sc^{3+}	0.87	+3.6%
Sr^{2+}	1.26	+50%
Y^{3+}	1.019	+21%
Yb^{3+}	1.125	+36%

From Shannon, R. D., Revised effective ionic radii and systematic studies of interatomic distances in halides and chalcogenides, *Acta Crystallogr.*, A32, 751, 1976. With permission.

1. Precipitated systems, in which the stabilizer has a very low solid solubility in the zirconia lattice, at temperatures where cation migration is still active, i.e., >1400K. Materials produced from such alloying additions are designated PSZ and are derived primarily from additions of MgO and CaO.
2. Solid solution systems where the stabilizer solubility is such that cation mobility has effectively ceased (or at least slowed down to be commercially practicable) and the stabilizer is retained in solid solution at relatively low temperatures. When formed, such systems are known as TZP and are produced by additions of Y_2O_3 and CeO_2.

It should also be remembered that two types of phase transformation and reactions are thus possible in zirconia systems, these being diffusional (generally occurring at temperatures >1400K) or diffusionless (military or martensitic, occurring <1400K) as discussed in Section III.

We will briefly examine the binary phase diagrams* of materials leading to both types of systems. We shall concentrate on those regions of the phase diagrams from which materials, with optimized transformation-toughening properties, may be derived.** While several phase diagrams have generally been proposed for each of the systems described, we shall concentrate only on the diagrams generally found to be the most accurate.

A. CaO-ZrO₂

While a number of workers[57-64] have examined the phase relations in the $CaO-ZrO_2$ system, confusion concerning the phases present, the actual composition, and the temperature of the eutectoid, still exist. The most recent diagram of the $ZrO_2-CaZrO_3$ region, although also still tentative, is presented Figure 18a. The main confusion in the diagram concerns the

---

* The term phase diagram is generally understood to refer to "phase equilibrium diagram." It should be remembered that ZTC materials used for their enhanced mechanical properties are rarely at equilibrium (see Sections IV.B and F).

** While a lot of effort has been expended in determining accurate phase equilibrium diagrams for the various systems (as mentioned in Section IV.A), optimized transformation-toughening materials are *not* at equilibrium.

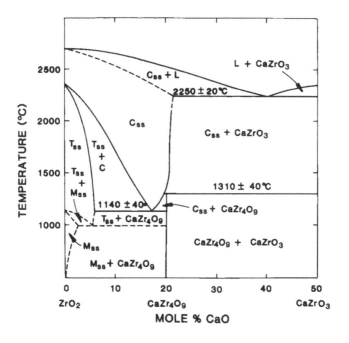

a

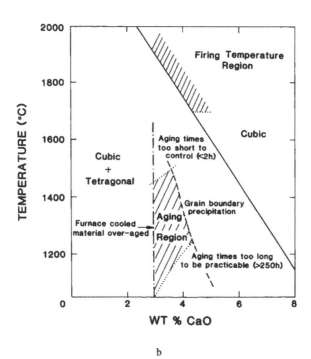

b

FIGURE 18. (a) Proposed equilibrium phase diagram for the ZrO_2-CaO system. (After Stubican, V. S. and Ray, S. P., *J. Am. Ceram. Soc.*, 60, 534, 1977.) (b) "Working" phase diagram for the ZrO_2-CaO system. The solubility of CaO in monoclinic and tetragonal is unknown, but small. (After Hannink, R. H. J., Johnston, K. A., Pascoe, R. T., and Garvie, R. C., *Science and Technology of Zirconia I*, 1981, 116.)

stability of the defect-fluorite ϕ_1-Ca Zr_4O_9 (20 mol% CaO) and ϕ_2-$Ca_6Zr_{19}O_{44}$ (24 mol % CaO, not shown in phase diagram) phases and the position of the cubic-eutectoid composition at 1140°C and 17 mol % CaO. Hangas et al.[65] have suggested that ϕ_1 may be metastable. For practical purposes, the cubic-eutectoid temperature is too low to affect the microstructures of these alloys for engineering applications.

From the diagram, it can be seen that the cubic-fluorite-stabilized-zirconia phase extends from about 10 to 20 mol % CaO at 1700°C. This cubic-stabilized-zirconia (CSZ) phase may be readily retained to room temperature by reasonably fast cooling.* While CSZ materials find applications as solid electrolytes, their mechanical properties are relatively poor. It was recognized that by lowering the alloy additions, improved materials could be obtained, so that PSZ was produced and a considerable improvement in mechanical properties could be achieved. Prior to 1975, PSZ materials consisted of monoclinic zirconia precipitates dispersed in a CSZ matrix. Present day methods involve cooling the materials sufficiently rapidly from the cubic phase field, suppressing the tetragonal to monoclinic transformation in any pre-cipitates and then bringing these precipitates to a state of metastability as described in Section III.

When fabricating engineering zirconia ceramics for commercial applications, a very limited composition and temperature regime is applicable. Figure 18b shows a typical "working" phase diagram of the CaO-ZrO_2 system.[66] Firing of an 8.4 mol % (4 wt %) CaO-ZrO_2 alloy consists of solution treating at 1800°C, followed by a "rapid" cool and a subsequent reheat to coarsen the previously formed tetragonal precipitates. The method of fabrication of pre-cipitated transformation-toughened zirconia systems is primarily the same for all systems. It is the accurate knowledge of the single phase cubic field and the eutectoid boundaries which are important for successful heat treatment programs in the fabrication process. We shall further discuss the thermal treatment of precipitated zirconia systems later (see Chapter 4).

B. MgO-ZrO_2

Diagrams for the MgO-ZrO_2 system have been presented by various workers.[1,57,67-69] The phase diagram proposed by Grain[69] has been found to be the most accurate for predicting phase behavior. The zirconia-rich end of the Grain diagram is shown in Figure 19.

At high temperatures, the MgO solubility in ZrO_2 is about 20 mol % at 2000°C, with the cubic phase being stable above 1400°C. At the eutectoid composition, 14 mol % (~5 wt %) MgO, the fully stabilized cubic form is readily retained to room temperature by suitable rapid cooling. Commercial PSZ materials are produced within the composition range 8 to 10 mol % (~2.8 to 3.5 wt %); the reasons for the limited composition range are similar to those depicted in the "working" diagram of the CaO-ZrO_2 system.

The main difference between the CaO-ZrO_2 and MgO-ZrO_2 diagrams is that the eutectoid temperature of the MgO-ZrO_2 system is well known, and occurs at 1400°C. Studies of the decomposition reaction have been performed by various workers. Duwez et al.[57] showed that the cubic solid solution decomposes into MgO and ZrO_2 constituents below 1400°C. Viechnicki and Stubican[68] have examined the decomposition kinetics as a function of MgO content in the range 12 to 20 mol % (~4.3 to 7.5 wt %), i.e., hypereutectoid compositions, and found that the kinetics were not significantly affected by alloy content, but markedly affected by the subeutectoid temperature. These workers observed that the maximum de-composition rate occured at 1200°C. Farmer et al.[70] examined the kinetics and microstructural relationships of the decomposed products in hypoeutectoid alloys in the range 8 to 11 mol % (~2.8 to 3.9 wt %) MgO. They concluded that the kinetics of the decomposition are

---

* It has been found, however, using transmission electron microscopy to examine a radiation/water-quenched 14 mol % CaO-ZrO_2 sample from 2000°C, that the formation of very small tetragonal precipitates within the CSZ matrix could not be suppressed.[106]

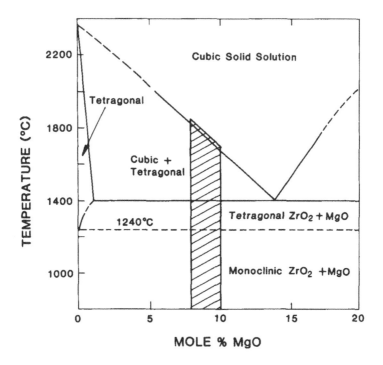

FIGURE 19. Zirconia-rich end of the MgO-ZrO$_2$ phase equilibrium diagram. (After Grain, C. F., *J. Am. Ceram. Soc.*, 50, 288, 1967.)

nucleation controlled, and microstructures produced may be categorized as a cellular reaction composed of rods growing perpendicular to the advancing reaction front.

All the investigations of the decomposition reaction agree with the predictions of the phase diagram (Figure 19), namely:

1. $\geq 1200°C$ Mg-CSZ $\rightarrow$ tetragonal ZrO$_2$ + MgO, on cooling tetragonal ZrO$_2$ $\rightarrow$ monoclinic ZrO$_2$
2. $\leq 1200°C$ Mg-CSZ $\rightarrow$ monoclinic ZrO$_2$ + MgO

An ordered compound, Mg$_2$Zr$_5$O$_{12}$ (not included in the phase diagram), was reported by Delamarre[71] to exist in stable equilbrium above 1800°C. In association with the decomposition reaction, the compound Mg$_2$Zr$_5$O$_{12}$ occurs as a metastable precursor to the decomposition reaction.[20,72-75] Hannink[72] has observed the occurrence of the compound in eutectoid and hypoeutectoid compositions as micro domains. These domains are stable <1400°C and may be induced in the range 1000 to 1300°C, after a suitable nucleation step <800°C. The Mg$_2$Zr$_5$O$_{12}$ prefers to nucleate and grow at the CSZ matrix-ZrO$_2$ precipitate interface, and is stable until the grain itself is consumed by the decomposition reaction. The effect of this phase on the mechanical properties is further discussed in Chapter 4.

A problem still exists in the phase diagram in that the precise solubility of MgO in monoclinic and tetragonal zirconia is not known with certainty. This arises because the spatial resolution of the instruments used to measure the MgO contents are not sufficiently accurate to determine MgO levels at low concentrations.

C. Y$_2$O$_3$-ZrO$_2$

This system has been extensively studied.[76-83] The main disagreements among all the diagrams is that of the eutectoid temperature. It is generally accepted, as a result of subsequent

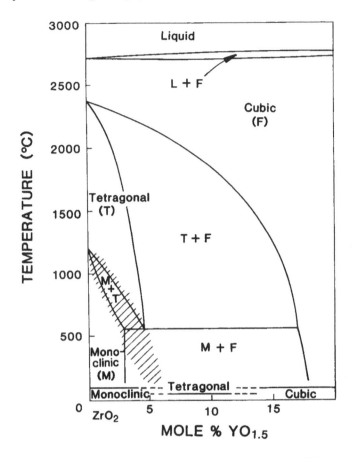

FIGURE 20.   Zirconia-rich end of the yttria-zirconia phase equilibrium diagram. Nonequilibrium homogeneous phases are indicated at the lower margin. Hatched region indicates nonequilibrium monoclinic-tetragonal transition. (After Scott, H. G., *J. Mater. Sci.*, 10, 1527, 1975.)

work, that the diagram of Scott[81] is the most accurate for ceramic processing purposes. The zirconia-rich end of Scott's diagram is shown in Figure 20. For fabrication purposes, the zirconia-rich end at temperatures above 1300°C is of most importance.

There are two essential differences between this system and the previously described CaO-ZrO_2 and MgO-ZrO_2 diagrams in terms of transformation-toughening zirconia ceramics. First, the extent of the tetragonal solid solution range is much larger in the yttria-based system, and second, the temperature of the tetragonal to monoclinic transformation is very low. The hatched region in Figure 20 indicates the nonequilibrium monoclinic transition.

Commercial transformation-toughening ceramics are generally produced within the composition range 2.5 to 3.5 mol % (~3 to 6 wt %) Y_2O_3-ZrO_2, by firing at 1400 to 1500°C followed by rapid cooling. This thermal treatment method forms the class of zirconia ceramics known as TZP. Precipitated PSZ materials can also be produced from composition in the range 3 to 6 mol % (~5 to 10 wt %) Y_2O_3-ZrO_2.

Because of the importance of the subsolidus (t)/(t-c) and (t-c)/(c) boundaries in the fabrication of commercial ceramics, Ruhle et al.[53] and Lanteri et al.[84] have independently attempted to determine the precise location of these boundaries. Their results, in relation to those of Scott,[81] are shown in Figure 21. The discrepancy is not large and has been accounted for in terms of the Cahn and Larche[85] model for coherent phase equilibrium.

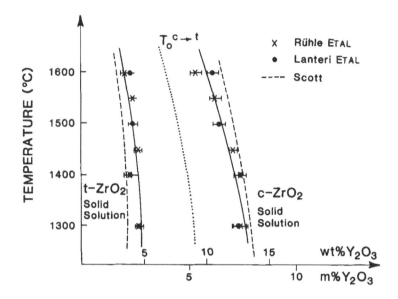

FIGURE 21. Zirconia-rich end of the yttria-zirconia phase equilibrium diagram. The equilibrium compositions were determined by energy dispersive X-ray spectroscopy. (Reference data from Ruhle et al.,[53] Lenteri et al.,[84] and Scott.[81] (After Ruhle, M., Claussen, N., and Heuer, A. H., *Science and Technology of Zirconia II*, 1984, 352.)

D. CeO$_2$-ZrO$_2$

The CeO$_2$-ZrO$_2$ system is another alloy system which shows considerable industrial potential. The zirconia-rich end of the Duwez and Odell[86] CeO$_2$-ZrO$_2$ phase diagram is shown in Figure 22. This system, like the yttria-zirconia system, also has an extensive solid-solution tetragonal phase field and has the monoclinic transformation temperature of a 20 mol% CeO$_2$-ZrO$_2$ alloy at about room temperature.

While modification concerning a eutectoid at 270°C has been proposed by Lange,[87] the phase field at the temperatures required for fabrication purpose, ~1400 to 1600°C, has not been disputed.

VI. RETENTION OF TETRAGONAL ZIRCONIA

Retention of the tetragonal phase is the most important factor for the utilization of the transformation-toughening phenomenon. As we have seen, under normal conditions of temperature and pressure, large particles of constrained pure tetragonal zirconia commence to transform to the monoclinic form at about 1200°C, the M$_s$ temperature. Transformation continues over a temperature range of about 100°C until it is finished at the M$_f$ temperature. We have seen in Sections IV.E and V that the M$_s$ temperature may be reduced, i.e., the tetragonal phase is retained by a number of mechanisms comprised essentially of chemical and physical effects.

Tetragonal particle/grain size retention (stability) has been addressed by a number of authors[37,87-93] using a number of approaches. As shown in Section IV.E, the stability of the tetragonal phase is most readily described in terms of the thermodynamics of the transformation. This approach determines the total free energy change associated with the transformation, of a spherical particle, by considering variables such as bulk and surface energy terms. We shall briefly review the retention of the tetragonal form for various conditions and situations of the zirconia particles. In this approach, we neglect the nucleation conditions

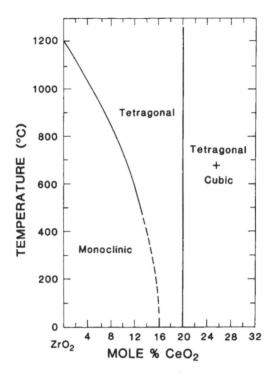

FIGURE 22.   Zirconia-rich end of the ZrO_2-CeO_2 phase
equilibrium diagram. (After Duwez, P. and Odell, P., *J.
Am. Ceram. Soc.*, 33, 274, 1950.)

(see Section IV.E) in order to give a simple explanation of the critical size effect associated
with retention of t-ZrO_2.

A. Unconstrained Particles

Very small "unconstrained" particles of tetragonal zirconia are known to exist in the free
state at room temperature. The small particle-size affect has been considered by Garvie[88]
using end-point-thermodynamic calculations. This approach yields the free energy, F, of a
particle as

$$F = (4/3)\pi \ r^3 F_{chem} + 4\pi r^2 S_{chem} \qquad (2)$$

where F_{chem} is the free energy/unit volume of a large crystal; r, the radius of a crystal under
consideration; and S_{chem}, surface energy of the crystal.

The difference in free energy between the tetragonal and monoclinic polymorphs is then
given by:

$$\Delta F_0 = (4/3)\pi r^3 (F_T - F_M) + 4\pi r^2 (S_T - S_M) \qquad (3)$$

where the subscripts T and M refer to the tetragonal and monoclinic polymorphs, respectively.
The tetragonal form can exist by considering a critical value of r_c when ΔF_0 is zero at a
particular temperature below the normal transformation temperature. Thus, we can write:

$$r_c = -3(F_T - F_M)/(S_T - S_M) \qquad (4)$$

Therefore, at or below some critical particle size, the bulk and surface energy balance

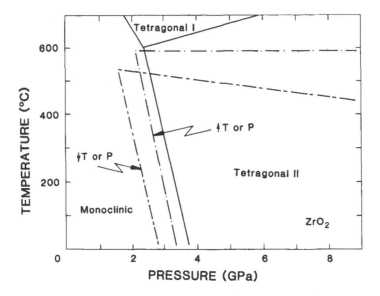

FIGURE 23. The pressure-temperature phase diagram of zirconia, showing hysteresis in the transformation of increasing temperature or pressure and decreasing temperature or pressure of previously transformed crystals. Pressure of 4 GPa at 200°C are required to initiate the transformation of virgin monoclinic zirconia crystals to the tetragonal form. (After Block, S., Da Jornada, J. H. A., and Piermarini, G. J., *J. Am. Ceram. Soc.*, 68, 497, 1985.)

allows room-temperature monoclinic particles to transform back to tetragonal. An experiment whereby monoclinic zirconia was ground continually finer was conducted by Bailey et al.[94,95] These workers observed a decrease in the monoclinic (111) X-ray lines and an increase in the (111) tetragonal lines when the particle size was about 10 nm.

Some reservations must be expressed about the validity of the experimental proof. Recently, Morgan[96] has produced precipitated monoclinic powders of ~6 nm diameter and Garvie[103] has made <10 nm zirconia powder of either form depending upon the pH of the solution from which precipitation occured. Therefore, absorbed species on the powder surface are known to play a very important part in determining the crystal form as a function of particle size.

B. Effects of Pressure and Temperature

The stability of the tetragonal phase as a function of hydrostatic pressure and temperature was first examined by Whitney.[97,98] By extrapolation of Whitney's pressure-temperature data, it could be shown that at a pressure of 3.9 GPa (29 bars), the tetragonal phase is retained at room temperature. Recently Block et al.[99] have reexamined the pressure-temperature phase relations of zirconia and found that, at room temperature, monoclinic zirconia transforms to tetragonal-II at 3.3 GPa on increasing pressure.* These workers also observed that the back transformation occurred at 2.785 GPa. Both forward and reverse transformations were affected by temperature. The modification by Block et al. of Bocquillon and Susses'[100] pressure-temperature phase diagram of zirconia is shown in Figure 23.

Block et al.[99] consistently observed that higher pressures were required to initiate the monoclinic to tetragonal transformation in pristine crystals, than in crystals which had been cycled through the transformation several times. This difference was attributed to an increased

---

* The tetragonal-II phase is related to the normally observed high temperature tetragonal-I form through the commonly referred to distorted fluorite structure.

number of potential transformation nucleation sites in the cycled crystals. Thus, we note how constraint, due to pure hydrostatic pressure, may retain the tetragonal phase and the importance of nucleating the transformation.

C. Effects of Matrix Constraint

For tetragonal zirconia particles, confined within a matrix, as in PSZ and ZTA, a consideration of bulk chemical and surface energy terms is not sufficient to account for the stability of the tetragonal phase.

Hannink et al.[66] have studied the influence of coarsening zirconia precipitates on the transformation temperature in Ca-PSZ. These workers showed that the reciprocal critical precipitate size was a linear function of temperature. Their approach could theoretically predict the transformation temperature if the total free energy change included bulk chemical and dilational strain energy plus changes in the particle-matrix interfacial and chemical free surface energy terms. It has since been shown by Lange,[87] Evans et al.,[90] and Lange and Green[91] and that in addition to the above terms, to obtain realistic values of strain and interfacial energies, energy contributions from twin boundaries and microcracking should also be considered.

Garvie and Swain[92] have expanded the approach of Hannink et al.[66] by including twinning and microcracking considerations where appropriate. By neglecting nucleation problems and using end-point thermodynamics, these workers evaluated different systems containing zirconia precipitates and particles, viz. Ca-PSZ, Al_2O_3-ZrO_2, and Mullite-ZrO_2. Expanding Equation 4 to include the additional terms, the free energy description for the transformation of a spherical constrained particle of radius r, in the absence of an applied stress becomes:

$$\Delta F_0 = (4/3)\pi r^3 (\Delta F_{CH} + \Delta F_D + \Delta F_{SH}) + 4\pi r^2(\Delta S_{CH} + \Delta S_{TW} + \Delta S_P) \qquad (5)$$

where ΔF_0 is the total free energy change, ΔF is the free energy change/unit volume, and the subscripts CH, D, and SH refer to the chemical, dilational, and shear energy contributions, respectively. ΔS denotes the free energy changes per unit area and the subscripts TW and P refer to the contributions from twin boundaries and precipitate (particle)/matrix interfaces, respectively. Summing the various terms, Equation 5 can be written more conveniently:

$$\Delta F_0/V = \Delta F_{CH} + \sum\Delta F_{ST} + (3/r) \sum\Delta S \qquad (6)$$

where V is the volume of the particle, $\Sigma\Delta F_{ST}$ the sum of the strain energy terms and $\Sigma\Delta S$ is the sum of the interfacial energy contributions. Expressing ΔF_{CH} in terms of experimental variables and applying conditions for equilibrium,[92] Equation 6 becomes:

$$1/r_c = \frac{q}{3\sum\Delta ST_b} T - \frac{q + \sum\Delta F_{ST}}{3\sum\Delta S} \qquad (7)$$

where r_c again denotes the critical particle size, q and T_b are the enthalpy of the transformation and transformation temperature of a crystal of "infinite" radius, and T is the transformation temperature (M_s) of a zirconia particle of radius r_c.

Inserting appropriate data into Equation 7 for Ca-PSZ, Al_2O_3, and Mullite-ZrO_2, Garvie and Swain[92] were able to show the linear temperature dependence of reciprocal radius for the tetragonal monoclinic transformation in these materials. Their calculations are compared with experimentally published data in Figure 24.

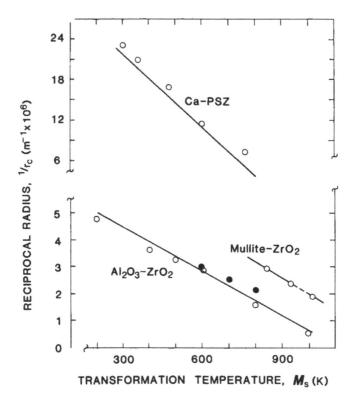

FIGURE 24. Temperature dependence of the reciprocal radius for the tetragonal to monoclinic transformation in Ca-PSZ, Al_2O_3-ZrO_2, and mullite-ZrO_2 materials. (Reference data for Ca-PSZ from Hannink et al.[66] Al_2O_3, from Claussen and Ruhle[101] and Heuer et al.,[37] mullite-ZrO_2, from Prochazha et al.[102]) (After Garvie, R. C. and Swain, M. V., *J. Mater. Sci.*, 20, 1193, 1985.)

REFERENCES

1. **Ryshkewitch, E.,** *Oxide Ceramics*, Academic Press, New York, 1960. 350.
2. **Garvie, R. C.,** Zirconium dioxide and some of its binary systems, *High Temperature Oxides, Part II,* Vol. 5, Alper, A. M., Ed., Academic Press, New York, 1970, 117.
3. **Subbarao, E. C., Maiti, H. S., and Srivastava, K. K.,** Martensitic transformation in zirconia, *Phys. Status Solidi A,* 21, 9, 1974.
4. **Yardley, K.,** The structure of baddeleyite and of prepared zirconia, *Mineral Mag.,* 2, 169, 1926.
5. **McCullough, J. D. and Trueblood, K. N.,** The crystal structure of baddeleyite (monoclinic ZrO_2), *Acta Crystallogr.,* 12, 507, 1955.
6. **Smith, D. K. and Newkirk, H. W.,** The crystal structure of baddeleyite (monoclinic ZrO_2) and its relation to the polymorphism of ZrO_2, *Acta Crystallogr.* 18, 983, 1965.
7. **Makovicky, E. and Hyde, B. C.,** Non-commensurate (misfit) layer structures, *Struct. Bonding (Berlin),* 46, 47, 1981.
8. **Sellar, J.,** Personal communication, 1985.
9. **Ruff, O. and Ebert, F.,** Refractory ceramics. I. The forms of zirconium dioxide, *Z. Anorg. Chem.,* 180, 19, 1929.
10. **Teufer, G.,** The crystal structure of tetragonal ZrO_2, *Acta Crystallogr.* 15, 1187, 1962.
11. **Barker, W. W., Bailey, F. P., and Garrett, W.,** A high temperature neutron diffraction study of pure and scandia-stabilized zirconia, *J. Solid State Chem.,* 7, 448, 1973.

12. **Smith, D. K. and Cline, C. F.**, Verification of existence of cubic zirconia at high temperature, *J. Am. Ceram. Soc.*, 45, 250, 1962.
13. **Wolten, G. M.**, Diffusionless phase transformations in zirconia and hafnia, *J. Am. Ceram. Soc.*, 43, 412, 1963.
14. **Wells, A. F.**, *Structural Inorganic Chemistry*, Oxford Press, London, 1962.
15. **Bendoraitis, J. G. and Salomon, R. E.**, Optical energy gaps in the monoclinic oxides of hanium and zirconium and their solid solutions, *J. Phys. Chem.*, 69, 3555, 1965.
16. **Martin, J. W. and Doherty, R. D.**, *Stability of Microstructures in Metallic Systems*, Cambridge University Press, Cambridge, 1980.
17. **Khachaturyan, A. G.**, *Theory of Structural Transformations in Solids*, John Wiley & Sons, New York, 1983.
18. **Cahn, J. W.**, Spinodal decomposition, *Trans. Metall. Soc. AIME*, 242, 167, 1968.
19. **Anderson, C. A. and Gupta, T. K.**, Phase stability and transformation toughening in zirconia, in *Science and Technology of Zirconia*, Advances in Ceramics, Vol. 3, Heuer, A. H. and Hobbs, L. W., Eds., American Ceramic Society, Columbus, Ohio, 1981, 184.
20. **Farmer, S. C., Mitchell, T. E., and Heuer, A. H.**, Diffusional decomposition of c-ZrO₂ in Mg-PSZ, in *Science and Technology of Zirconia II*, Advances in Ceramics Vol. 12, Ruhle, M., Clussen, N., and Heuer, A. H., Eds., American Ceramic Soiety, Columbus, Ohio, 1983, 152.
21. **Cohen, M. and Kaufman, L.**, Martensitic transformations, in *Progress in Metal Physics 7*, Pergamon Press, London, 1958.
22. **Mackenzie, J. K.**, The crystallography of martensite transformation, *J. Aust. Inst. Met.*, 5, 90, 1960.
23. **Christian, J. W.**, *The Theory of Transformations in Metals*, Pergamon Press, Oxford, 1965.
24. **Wayman, C. M.**, *Introduction to the Crystallography of Martensitic Transformations* Macmillan, New York, 1965.
25. **Rotburd, A. L.**, Martensitic transformation as a typical phase transformation in solids, in *Solid State Physics, Vol. 33*, Academic Press, New York, 1978, 317.
26. **Bain, E. C.**, Nature of Martensite, *Trans. Metall. Soc. AIME*, 70, 25, 1924.
27. **Greninger, A. B. and Troiano, A. R.**, The mechanism of martensite formation, *Trans. Metall. Soc. AIME*, 185, 590, 1949.
28. **Wechsler, M. S., Lieberman, D. S., and Read, T. A.**, On the theory of the formation of martensite, *Trans. Metall. Soc. AIME*, 197, 1503, 1953.
29. **Bowles, J. S. and Mackenzie, J. K.**, The crystallography of martensite Transformations. I, *Acta Metall.*, 2, 129, 1954; **Bowles, J. S. and Mackenzie, J. K.**, The crystallography of martensite transformation. II, *Acta Metall.*, 2, 138, 1954; **Bowles, J. S. and Mackenzie, J. K.**, The Crystallography of Martensite Transformation. III. Face-centred to Body-centred Tetragonal Transformations, *Acta Metall.*, 2, 224, 1954.
30. **Kriven, W. M.**, Shear transformations in inorganic materials, in *An International Conference on Solid-Solid Phase Transformations*, Aaronson, H. I., Laughlin, D. E., Sekerka, R. F., and Wayman, C. M., Eds., Carnegie-Mellon University, Pittsburgh, 1982, 1507.
31. **Kennedy, S. W.**, Structural correspondences and mechanisms for the polymorphic transformation NaCl (CsCl type) and NaCl (rhombohedral), *J. Solid State Chem.*, 34, 31, 1980.
32. **Bailey, J. E.**, The monoclinic to tetragonal transformation and associated twinning in thin films of zirconia, *Proc. R. Soc. London Ser. A*, 279, 396, 1964.
33. **Bansal, G. K. and Heuer, A. H.**, Martensitic phase transformation in zirconia (ZrO₂). I. metallographic evidence, *Acta Metall.*, 20, 1281, 1972.
34. **Bansal, G. K. and Heuer, A. H.**, Martensitic phase transformation in zirconia (ZrO₂). II. crystallographic aspects, *Acta Metall.*, 22, 409, 1972.
35. **Chen, I.-W. and Chiao, Y.-H.**, Theory and experiment of martensitic nucleation of ZrO₂ containing ceramics and ferrous alloys, *Acta Metall.*, 33, 1827, 1985; **Chen, I.-W., Chiao, Y.-H., and Tsuzaki, K.**, Statistics of martensitic nucleation, *Acta Metall.*, 33, 1847, 1985.
36. **Heuer, A. H. and Ruhle, M.**, On the nucleation of the martensitic transformation in zirconia (ZrO₂), *Acta Metall.*, 33, 2101, 1985.
37. **Heuer, A. H., Claussen, N., Kriven, W. M., and Ruhle, M.**, Stability of tetragonal ZrO₂ particles in ceramic matrices, *J. Am. Ceram. Soc.*, 65, 642, 1982.
38. **Guenin, G. and Gobin, P. F.**, A localized soft mode model for the nucleation of thermoelastic martensitic transformation: application to the 9R transformation, *Metall. Trans.*, 13A, 1127, 1982.
39. **Clapp, P. C.**, A localized soft mode theory for martensitic transformations, *Phys. Status Solidi B* 57, 561, 1973.
40. **Garvie, R. C. and Burke, S.**, Soft phonon modes and the monoclinic tetragonal transformation in zirconia and hafnia, *J. Mater. Sci. Lett.*, 12, 1487, 1977.
41. **Muddle, B. C. and Hannink, R. H. J.**, Crystallography of the tetragonal to monoclinic transformation in MgO-partially stabilized zirconia, *J. Am. Ceram. Soc.*, 69, 547, 1986.

42. **Lenz, L. K. and Heuer, A. H.**, Stress-induced transformation during subcritical crack growth in partially stabilized zirconia, *J. Am. Ceram. Soc.*, 65, C192, 1982.

43. **Schoenlein, L. H. and Heuer, A. H.**, Transformation zones in MgO-PSZ, in *Fracture Mechanics of Ceramics*, Vol. 6, Bradt, R. C., Evans, A. G., Hasselman, D. P. H., and Lange, F. F., Eds., Plenum Press, New York, 1983, 309.

44. **Heuer, A. H., Schoenlein, L. H., and Farmer, S.**, New Microstructural Features in Magnesia Partially Stabilized Zirconia (Mg-PSZ), in *Science and Ceramics*, Vol. 12, Vincenzini, P., Ed., Ceramurgica s.r.l., Faenza, Italy, 1983, 257.

45. **Muddle, B. C. and Hannink, R. H. J.**, Phase transformations involving an orthorhombic phase in MgO-partially stabilized zirconia, in *Science and Technology of Zirconia III, Advances in Ceramics*, Vol. 24, Tokyo, Japan, 1986, in press.

46. **Hannink, R. H. J., Porter, J. D., and Marshall, D. B.**, Direct observation of cycled phase transformation in zirconia, *J. Am. Ceram. Soc.*, 69, C116, 1986.

47. **Kelly, P. M. and Ball, C. J.**, Crystallography of stress-induced martensitic transformations in partially stabilized zirconia, *J. Am. Ceram. Soc.*, 69, 259, 1986.

48. **Hannink, R. H. J. and Swain, M. V.**, A mode of deformation in partially stabilized zirconia, *J. Mater. Sci.*, 16, 1428, 1981.

49. **Kriven, W. M.**, The transformation mechanism of spherical zirconia particles in alumina, in *Science and Technology of Zirconia II*, Claussen, N., Ruhle, M., and Heuer, A. H., Eds., The American Ceramic Society, Columbus, Ohio, 1984, 64.

50. **Acton, A. F., Bevis, M., Crocker, A. G., and Ross, N. H.**, Transformation strains in lattices, *Proc. R. Soc. London Ser. A*, 320, 101, 1970.

51. **Crocker, A. G.**, The phenomenological theories of martensite crystallography, *J. Phys. (Paris) Colloq.*, (C4), 209, 1982.

52. **Bischoff, E. and Ruhle, M.**, Twin boundaries in monoclinic ZrO_2 particles confined in a mullite matrix, *J. Am. Ceram. Soc.* 66, 123, 1983.

53. **Ruhle, M., Claussen, N., and Heuer, A. H.**, Microstructural studies of Y_2O_3 containing tetragonal ZrO_2 polycrystals (Y-TZP), in *Science and Technology of Zirconia II*, Ruhle, M., Claussen, N., and Heuer, A. H., Eds., The American Ceramic Society, Columbus, Ohio, 1984, 352.

54. **Ruhle, M., Kraus, B., Strecker, A., and Waidelich, D.**, *In-situ* observations of stress-induced phase transformations in ZrO_2-containing ceramics, in *Science and Technology of Zirconia II*, Claussen, N., Ruhle, M., and Heuer, A. H., Eds., The American Ceramic Society, Columbus, Ohio, 1984, 256.

55. **Hannink, R. H. J., Muddle, B. C., and Swain, M. V.**, Transformation plasticity in tetragonal zirconia polycrystals, in *Austceram '86 Proceedings*, The Australian Ceramic Society, Melbourne, Australia, 1986, 145.

56. **Barker, W. W. and Williams, L. S.**, Some limitations of cubic stabilization in zirconia, *J. Am. Ceram. Soc.*, 4, 1, 1968.

57. **Duwez, P., Odell, F., and Brown, F. H.**, Stabilization of zirconia with calcia and magnesia, *J. Am. Ceram. Soc.*, 35, 107, 1952.

58. **Dietzel, A. and Tober, H.**, Uber Zirkonoxyd und Zweistoff Systeme mit Zirkonoxyd, *Ber. Dtsch. Keram. Ges.*, 30, 71, 1953.

59. **Nadler, M. R. and Fitzsimmon, E. S.**, Preparation and properties of calcium zirconate, *J. Am. Ceram. Soc.*, 38, 214, 1955.

60. **Cocco, A.**, Composition limits at high temperatures of the cubic phase composed of ZrO_2 and CaO, *Chim. Ind. (Milan)*, 41, 882, 1959.

61. **Tien, T. Y. and Subbarao, E. C.**, X-ray and electrical conductivity study of the fluorite phase in the system ZrO_2-CaO, *J. Chem. Phys.*, 39, 1041, 1963.

62. **Garvie, R. C.**, The cubic field in the system CaO-ZrO_2, *J. Am. Ceram. Soc.*, 51, 553, 1968.

63. **Stubican, V. S. and Ray, S. P.**, Phase equilibria and ordering in the system ZrO_2-CaO, *J. Am. Ceram. Soc.*, 60, 534, 1977.

64. **Hellman, J. R. and Stubican, V. S.**, Stable and metastable phase relations in the system ZrO_2-CaO, *J. Am. Ceram. Soc.*, 66, 260, 1983.

65. **Hangas, J., Mitchell, T. E., and Heuer, A. H.**, Ordered compounds in the system CaO-ZrO_2, in *Science and Technology of Zirconia II*, Claussen, N., Ruhle, M., and Heuer, A. H., Eds., The American Ceramic Society, Columbus, Ohio, 1984, 107.

66. **Hannink, R. H. J., Johnston, K. A., Pascoe, R. T., and Garvie, R. C.**, Microstructural changes during isothermal aging of a calcia partially stabilized zirconia alloy, in *Science and Technology of Zirconia I*, Heuer, A. H. and Hobbs, L. W., Eds., The American Ceramic Society, Columbus, Ohio, 1981, 116.

67. **Ebert, F. and Cohn, E.**, The ceramics of highly refractory materials. VI. The system: ZrO_2-MgO, *Z. Anorg. Allg. Chem.*, 213, 321, 1933.

68. **Viechnicki, D. and Stubican, V. S.**, Mechanisms of decomposition of the cubic solid solutions in the system ZrO_2-MgO, *J. Am. Ceram. Soc.*, 48, 292, 1965.

69. **Grain, C. F.**, Phase relations in the ZrO₂-MgO system, *J. Am. Ceram. Soc.*, 50, 288, 1967.

70. **Farmer, S. C., Heuer, A. H., and Hannink, R. H. J.**, Eutectoid decomposition of MgO-partially stabilized ZrO₂, *J. Am. Ceram. Soc.*, 70, 431, 1987.

71. **Delamarre, C.**, Contributions a l'etude de quelques systemes HfO₂-MO: comparison avec les systemes correspondants a base de zircone, *Rev. Int. Hautes Temp. Refract.*, 9, 209, 1972.

72. **Hannink, R. H. J.**, Microstructural development of sub-eutectoid aged MgO-ZrO₂ alloys, *J. Mater. Sci.*, 18, 457, 1983.

73. **Hannink, R. H. J. and Rossell, H. J.**, An electron-optical study of the cubic MgO-stabilized zirconia decomposition, *Micron*, Suppl. 11, 36, 1980.

74. **Rossell, H. J. and Hannink, R. H. J.**, The phase Mg₂Zr₅O₁₂ in MgO partially stabilized zirconia, in *Science and Technology of Ziconia II*, Ruhle, M., Claussen, N., and Heuer, A. H., The American Ceramic Society, Columbus, Ohio, pp 139.

75. **Chaim, R. and Brandon, D. G.**, Microstructural evolution and ordering in commercial Mg-PSZ, *J. Mater. Sci.*, 19, 2934, 1984.

76. **Duwez, P., Brown, F. H., and Odell, F.**, The zirconia-yttria System, *J. Electrochem. Soc.*, 98, 356, 1951.

77. **Fan, F.-K., Kuznetsov, A. K., and Keler, E. K.**, Phase relations in Y₂O₃-ZrO₂ system, *Izv. Akad. Nauk SSSR Ser. Khim.*, 1, 1141, 1962; **Fan, F.-K., Kuznetsov, A. K., and Keler, E. K.**, Phase relations in Y₂O₃-ZrO₂ system, *Izv. Akad. Nauk SSSR Ser. Khim.*, 2, 601, 1963.

78. **Lefevre, J.**, Some structural modifications of fluorite-type phases in systems based on zirconia or hanium oxide, *Ann. Chim. (Paris)*, 8, 117, 1963.

79. **Smith, D. K.**, The non-existence of yttrium zirconia, *J. Am. Ceram. Soc.*, 49, 625, 1966.

80. **Srivastava, K. K., Patil, R. N., Choudhary, C. B., Gokhale, K. V., and Subbarao, E. C.**, Revised phase diagram of the system zirconia-yttrium oxide (ZrO₂-YO₁.₅), *Trans. J. Br. Ceram. Soc.*, 73, 85, 1974.

81. **Scott, H. G.**, Phase relationships in the zirconia-yttria system, *J. Mater. Sci.* 10, 1527, 1975.

82. **Stubican, V. S., Hink, R. C., and Ray, S. P.**, Phase equilibria and ordering in the system ZrO₂-Y₂O₃, *J. Am. Ceram. Soc.*, 61, 18, 1978.

83. **Pascual, C. and Duran, P.**, Subsolidus phase equilibria and ordering in the system ZrO₂-Y₂O₃, *J. Am. Ceram. Soc.*, 66, 23, 1983.

84. **Lanteri, V., Heuer, A. H., and Mitchell, T. E.**, Tetragonal phase in the system ZrO₂-Y₂O₃, in *Science and Technology of Zirconia II*, Ruhle, M., Clussen, N., and Heuer, A. H., Eds., The American Ceramic Society, Columbus, Ohio, 1984, 118.

85. **Cahn, J. W. and Larche, F.**, A simple model for coherent equilibrium, *Acta Metall.*, 32, 1915, 1984.

86. **Duwez, P. and Odell, P.**, Phase relationships in the system zirconia-ceria, *J. Am. Ceram. Soc.*, 33, 274, 1950.

87. **Lange, F. F.**, Transformation toughening. I — V, *J. Mater. Sci.*, 17, 225, 1982.

88. **Garvie, R. C.**, The occurence of metastable tetragonal zirconia as a crystallite size effect, *J. Phys. Chem.*, 69, 1238, 1965.

89. **Garvie, R. C.**, Stability of the tetragonal structure in zirconia microcrystals, *J. Phys. Chem.*, 82, 218, 1978.

90. **Evans, A. G., Burlingame, N., Drory, M., and Kriven, W. M.**, Martensitic transformations in zirconia-particle size effects and toughening, *Acta Metall.*, 29, 447, 1981.

91. **Lange, F. F. and Green, D. J.**, Effect of inclusion size on the retention of tetragonal ZrO₂: theory and experiments, in *Science and Technology of Zirconia I*, Heuer, A. H., and Hobbs, L. W., Eds., The American Ceramic Society, Ohio, 1981, 217.

92. **Garvie, R. C. and Swain, M. V.**, Thermodynamics of the tetragonal to monoclinic phase transformation in constrained zirconia microcrystals. I, *J. Mater. Sci.*, 20, 1193, 1985.

93. **Garvie, R. C.**, Thermodynamics of the tetragonal to monoclinic phase transformation in constrained zirconia microcrystals. II, *J. Mater. Sci.*, 20, 3479, 1985.

94. **Bailey, J. E., Lewis, D., Librant, Z. M., and Porter, L. J.**, Phase transformations in milled zirconia, *Trans. J. Br. Ceram. Soc.*, 71, 25, 1972.

95. **Bailey, J. E., Bills, P. M., and Lewis, D.**, Phase stability in hafnium oxide powders, *Trans. J. Br. Ceram. Soc.*, 74, 247, 1975.

96. **Morgan, P. E. D.**, Synthesis of 6 nm ultrafine monoclinic zirconia, *J. Am. Ceram. Soc.*, 67, C204, 1984.

97. **Whitney, E. D.**, Effect of pressure on monoclinic-tetragonal transition of zirconia; thermodynamics, *J. Am. Ceram. Soc.*, 45, 612, 1962.

98. **Whitney, E. D.**, Electrical resistivity and diffusionless phase transformations of zirconia of high temperatures and ultrahigh pressures, *J. Electrochem. Soc.*, 112, 91, 1965.

99. **Block, S., Da Jornada, J. H. A., and Piermarini, G. J.**, Pressure temperature diagram of zirconia, *J. Am. Ceram. Soc.*, 68, 497, 1985.

100. **Bocquillon, G. and Susse, C.**, Diagramme de phase de la zircone sous pression, *Rev. Int. Hautes Temp. Refract.*, 6, 263, 1969.

101. **Claussen, N. and Ruhle, M.** Design of transformation-toughened ceramics, in *Science and Technology of Zirconia I*, Heuer, A. H. and Hobbs, L. W., Eds., The American Ceramic Society, Columbus, Ohio, 1981, 137.
102. **Prochazka, S., Wallace, J. S., and Claussen, N.**, Microstructure of sintered mullite-zirconia composites. *J. Am. Ceram. Soc.*, 66, C125, 1983.
103. **Garvie, R. C.**, Unpublished results.
104. **Adam, J. and Rogers, M. D.**, The crystal structure of ZrO_2 and HfO_2, *Acta Crystallogr.*, 12, 951, 1959.
105. **Scott, H. G.**, Unpublished results.
106. **Hannink, R. H. J.**, Unpublished data.

CHAPTER 3

MECHANICS AND MECHANISMS OF TOUGHENING

I. INTRODUCTION

In the last decade or so, it has been established that materials containing t-ZrO_2 that undergo a stress-induced phase transformation during fracture can have strengths and fracture-toughness values much higher than most other ceramics. Such large increases in toughness are unprecedented in materials that are incapable of crack-tip blunting. Remarkably, the science involved in the understanding of this phenomenon has kept pace with the technological developments. It is particularly important to understand the scientific principles that underlie the cause and magnitude of the toughening in such materials, since these will be of key importance in the reliable use of these materials and in the development of new materials. The aim of this chapter is first to introduce the critical concepts that are required to understand the fracture of brittle solids, and second to discuss the mechanics that are required to understand the mechanisms and predict the amount of toughening in a material that undergoes a stress-induced transformation. In some transformation-toughened systems, other toughening mechanisms, such as the formation of microcracks and the deflection of cracks by second-phase obstacles, appear to play a role and these will be briefly discussed at the end of the chapter.

II. FRACTURE MECHANICS CONCEPTS

Fracture has been of importance to humankind since the shaping of primitive, Stone Age tools was first introduced, and yet the scientific understanding of this subject has only been developed during this century. The work by Griffith[1] is generally considered the first breakthrough in developing a sound, scientific basis for fracture. He reasoned that the "free energy" of a cracked body under stress should *not* increase during crack extension. Consider an infinite plate of unit thickness that contains a through-the-thickness crack of length 2c and is being subjected to a uniform tensile stress (σ), as shown in Figure 1. The total energy (U) of the system may be written as:

$$U = U_0 + U_E - W + U_s \tag{1}$$

where U_0 is the elastic energy of the uncracked loaded plate (a constant), U_E is the change in elastic energy caused by the introduction of the crack, W is the work performed by the external forces and U_s is the energy associated with the formation of new surfaces. Utilizing an elastic solution put forward by Inglis,[2] Griffith was able to show for an atomically sharp crack, which is expected in brittle materials, that

$$(U_E - W) = -\pi\sigma^2c^2/E \tag{2}$$

where E is the Young's modulus of the material. For equilibrium conditions, the term U_s is given by the product of the crack area, the number of surfaces formed (two) and the thermodynamic surface energy (γ), i.e., 4 cγ for a plate of unit thickness. If the total energy of the system is not to increase during crack extension, the condition for failure is simply given by dU/dc $\leq$ 0. If we apply the failure criterion to Equation 1, we obtain

$$dU/dc = \{d(U_E - W)/dc\} + \{dU_s/dc\} \leq 0 \tag{3}$$

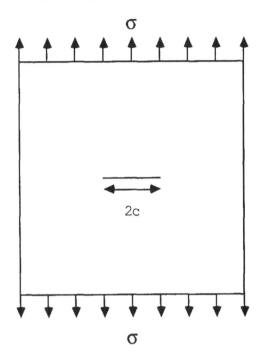

FIGURE 1.   An infinite plate loaded in uniform tension (σ) that contains a "through-the-thickness" crack.

Substituting the various terms into Equation 3 and rearranging we can obtain the critical stress for crack extension (the fracture stress, σ_f), i.e.,

$$\sigma_f \geq (2E\gamma/\pi c)^{1/2} \tag{4}$$

If one considers U as a function of c, it is found that once Equation 4 is satisfied, that U decreases with further increases in c, implying the crack growth will be unstable. Equation 4 shows that the fracture stress depends on the "material parameters" E and γ and the "crack sizes" c, that are assumed to be present in a material.

For ceramic materials, it is now generally considered that there are a large number of possible microstructural sources for these cracks and thus one considers the conditions that lead to a particular type of flaw population. For brittle materials, the cracks are generally assumed to form by cleavage of atomic bonds in regions in which there are high stresses. These stresses may be due to stress concentrations or residual strains and they will be particularly effective in producing cracks if weak interfaces, that are easy to cleave, are available. The high stresses are generally associated with the heterogeneous nature of the material at the microstructural level or inelastic deformation that cannot be accommodated. For example, voids may be present in a material as a result of the processing, and, during service, the stress concentration at a sharp corner of these voids may be sufficient to produce a crack. Thus, cracks can be a result of the processed microstructure and may be present before use, or they may form during subsequent service. In this latter category, it is known that high stresses can occur in contact events (e.g., impact, erosion, wear, etc.), leading to crack formation in the vicinity of the contact site. In addition to having a variety of "flaw populations" that compete to be fracture origins, the crack sizes within any one population will form a distribution and thus the fracture stress of a brittle material is best considered as a distribution rather than a fixed number. The complexity of failure sources in brittle materials has led to empirical statistical approaches for describing "strength distributions."

After Griffith's work, it became clear that the theory only really applied to ideally brittle materials. For example, in metals the energy to form fracture surfaces must include a large plastic energy term. Even in ceramics, attempts to measure "γ" from fracture tests have shown its magnitude is generally much higher than the thermodynamic surface energy. The implication is that there are other energy-dissipative mechanisms occurring when one forms a fracture surface in a ceramic material. These may include acoustic emission, heat, inelastic deformation, or microstructural interactions. It is these latter two groups that have been of particular interest to ceramic scientists as the implication is that, if they can be controlled, one should be able to make it more difficult for a crack to propagate, In any case, it was clear that for the Griffith concept to be useful, it was in need of modification.

In the period following World War II, Irwin suggested that one could consider the energy balance of Equation 1 in a slightly different way.[3] In Equation 3, the first two terms involving W and U_E were coupled together because they involve terms that act to promote crack extension, whereas the second term involving U_S represents the resistance of the material. One can, therefore, define a parameter $G = -d(U_E - W)/dA$, which represents the elastic energy per unit crack area (A) that is available for infinitesimal crack extension. G has been given a number of names in the literature; in this work it will be termed the "crack extension force." The units of G are in terms of an energy per unit area, but this can also be thought of as a force per unit crack length. In terms of the crack geometry in Figure 1 and using $dA = 2dc$ for a plate of unit thickness with Equation 2, the value of G is given by

$$G = -d(U_E - W)/[2dc] = \pi\sigma^2c/E \qquad (5)$$

Crack extension occurs when G reaches a critical value (G_C) which is equal to dU_S/dA and this latter term now contains all the dissipative terms that are characteristic of the formation of fracture surface in a particular material. This term represents the crack resistance processes available in a material and is called the "crack resistance force," R. Thus, crack extension occurs when the crack extension force reaches a critical value and is equal to the crack resistance force, i.e., $G = G_C = R$. In the ceramics lieterature, it has been popular to keep the formalism similar to Griffith and instead of R, a term γ_f has been used to replace γ. This is generally termed the "fracture surface energy", where $U_S = 4c\gamma_f$ and thus $R = dU_S/[2dc] = 2\gamma_f$. Failure for the geometry considered by Griffith is now given, using Equation 5, as $G_C = \pi\sigma_f^2c/E = R$. The interplay between G and R is shown in Figure 2. In this figure, the failure condition is when the G and R curves intersect and the value of G at this point is G_C. We have assumed here that R is independent of crack length. Consider in Figure 2 that the body contains a crack length c_1. If we apply a stress σ_1, the $G(\sigma_1) < R$ for this crack length and the crack will not extend. If, however, we apply a stress σ_2, then $G = R$ for this crack length and the crack will propagate. Moreover, as the crack propagates and increases in length, G increases further with respect to R and the crack propagation is thus "unstable". Another way of considering this diagram is to say for a stress σ_2, the "critical crack length" is c_1, or at this stress, c_1 is the maximum size of crack the body can tolerate. Increasing R, therefore, increases the critical crack size, i.e., increases the size of cracks that a body can withstand at a given stress (or a given G).

At the same time as the G concept was being developed, it became clear to Irwin[3] that an alternative approach could be used to describe fracture. This alternative was based on the idea that one could solve a variety of elastic problems involving cracks. All stress systems in the vicinity of a crack can be derived from three modes of loading and these are illustrated in Figure 3. In all these linear elastic solutions, it was found that the stresses in the vicinity of a crack tip (σ_{ij}) take the form[4]

$$\sigma_{ij} = K f_{ij}(\Theta)/[2\pi r]^{1/2} \qquad (6)$$

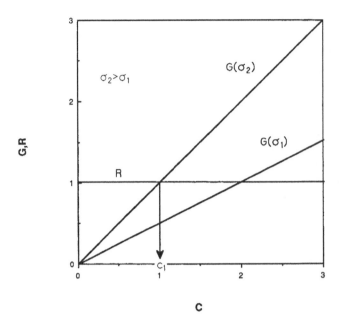

FIGURE 2. The crack extension force (G) for two different applied stresses and the crack resistance force (R) as a function of crack length for the loading geometry depicted in Figure 1. At crack length, c_1, the crack will propagate (G = R) for an applied stress σ_2.

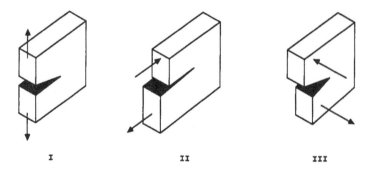

FIGURE 3. The three modes of crack loading: crack opening (Mode I), in-plane shear (Mode II), and tearing (Mode III).

where r and Θ are cylindrical polar coordinates of a point with respect to the crack tip (Figure 4), K is a parameter called the "stress intensity factor" and $f_{ij}(\Theta)$ represents the angular variation of the stresses. The parameter K is thus a scaling factor for the magnitude of the stresses near a crack tip or the amplitude of the elastic field. Equation 6 shows only the leading term of a series expansion from the general elastic solution, but it is the largest term for the region near a sharp crack tip. K is determined by the boundary conditions of the elastic problem and it can be shown for a given mode that $K = \sigma Y(c)^{1/2}$, where Y is a dimensionless parameter that depends on the geometries of the crack and the specimen and σ is the *applied* stress. Y values have been tabulated for many different types of geometries and are often available in handbooks.[5,6] For example, in the geometry considered by Griffith (Figure 1) $Y = \pi^{1/2}$. One might expect that K, or the magnitude of the crack tip stresses, would be related to the crack extension force G and Irwin was able to show that there was

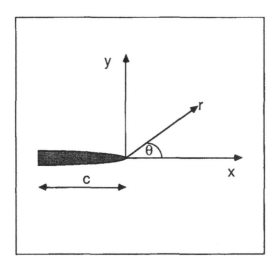

FIGURE 4. Cylindrical polar (r,Θ) and cartesian coordinates (x,y) used to describe region around a crack tip.

a simple relationship between G and K.[3] Ceramics generally fail and are tested in Mode I, and in this situation, the relationship between G and K is given by $G = (K_I^2/E)$ and $G = (K_I^2[1 - v^2]/E)$ for plane stress and plane strain conditions, respectively. The subscript on K is to identify the mode of loading and v is Poisson's ratio. The key point is that a fracture criterion of $G = G_C$ is equivalent to a critical value of K for a given mode. For a pure Mode I fracture, the criterion for crack extension is given by $K_I = K_{IC}$. In recent times, K_{IC} has often been called the "fracture toughness" of a material, whereas this term had previously been used for G_C. If we return now to the geometry considered in Figure 1, we find the failure condition can be described in a variety of ways, e.g.,

$$G_C = R = 2\gamma_f = Y^2\sigma_f^2 c = K_{IC}^2/E$$

or

$$K_{IC} = (EG_C)^{1/2} = (2E\gamma_f)^{1/2} = \sigma_f Y(c)^{1/2} \tag{7}$$

where $Y = (\pi)^{1/2}$. For a more detailed discussion of the fracture behaviour of brittle solids, the reader is referred to the book by Lawn and Wilshaw.[7]

In order to determine K_{IC} of a material, the general approach is to introduce a crack of known size into a material and measure its strength. Thus, as long as we know Y for the test geometry, we can determine K_{IC} using $K_{IC} = \sigma_f Y(c)^{1/2}$ (Equation 7). Using the type of formalism we have outlined, once we know the value of K_{IC} for a material, then we know the size of flaw it can tolerate at a given stress. As we shall see in Chapter 6, another important aspect of the fracture toughness of a material is that in situations where new flaws are created, such as in contact damage, the higher the toughness of a material, the more difficult it is to create flaws of a given size. Like the Griffith approach, Equation 7 indicates that strength depends on a combination of a material property (K_{IC}) and flaw size. In the Irwin approach, however, the magnitude of K_{IC} is expected to have a component that depends on the microstructure of the material. Thus, if we can identify mechanisms that increase K_{IC} for a material, we have an approach to increasing the strength of a material, or the critical flaw size, that it can withstand at a given stress. It is this approach that has produced new ceramic materials in recent years with higher strengths and toughness values than

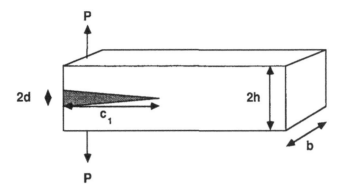

FIGURE 5.   The double cantilever beam (DCB) crack geometry containing a crack (length c_1) under applied load (P). The opening of the crack is shown schematically to exaggerate its opening displacement (2d).

previously considered possible. The subject of this book, transformation toughening, is an example of such a mechanism for increasing K_{IC}. Toughening mechanisms in ceramics have been identified to be of three distinct types: crack tip shielding, crack tip interactions, and crack bridging. Crack tip shielding involves the reduction of the stresses in the crack tip region below those expected from the applied stress intensity factor. Transformation and microcrack toughening are examples in ceramics of such mechanisms and these will be discussed in detail in this chapter. Crack tip interaction occurs when the crack interacts with obstacles that impede or deflect the motion of the crack. Examples of these obstacles are second-phase particles, whiskers, or fibers. Zirconia-toughened ceramics (ZTC) are generally multiphase and, as such, crack tip interactions are expected to play a role. These will also be discussed briefly later in this chapter. Crack-bridging mechanisms involve obstacles being left behind the advancing crack front, which connect the crack faces together. These bridges or ligaments restrain the opening of the crack faces, thereby increasing toughness.

It is worthwhile at this point to consider some more-subtle details of crack motion at the critical condition for crack extension. The value of G for the geometry described in Figure 1 is linearly proportional to crack length (Equation 5). For other geometries this may not be the case. Consider the double cantilever beam (DCB) crack geometry shown in Figure 5. Using a simple approach, it is possible to show that the crack extension depends on the way the load is applied. For example, for a constant load test, the crack extension force is $G(P) \sim P^2c^2/[EIb]$, where I is the moment of inertia of the beams and b is their depth.[8] For a fixed grip test, in which one keeps the displacement of the beam ends (d) a constant, the crack extension force, $G(d) \sim 9d^2EI/[bc^4]$.[8] Figure 6 compares these two types of loading for the critical condition for crack extension. For the constant load test, we see that G(P) increases as the crack propagates, whereas G(d) decreases. The implication is that for a constant load test once G = R, "unstable crack growth" occurs (G>R). For the fixed grip test, however, once the crack propagates, it will immediately arrest as G < R. In this latter case, d needs to be increased further for the crack to extend, thus "stable crack growth" occurs. For unstable crack growth, the requirement is that G = R *and* dG/dc>dR/dc.

A final important subject is our previous assumption that R is independent of crack length in Figures 2 and 6. This is not necessarily the case and, in some cases, the energy dissipation rate by the material can increase as the crack extends. Thus, the crack resistance force, R, in some materials increases with increasing crack length. As we shall see, this effect occurs in ZTC. Consider a hypothetical R curve shown in Figure 7 for a material loaded as in Figure 1. Here we see a complex crack growth behavior. If the material contains a crack of length c_1, the value of $G(\sigma_1)$ is such that unstable crack extension occurs until $G(\sigma_1) <$

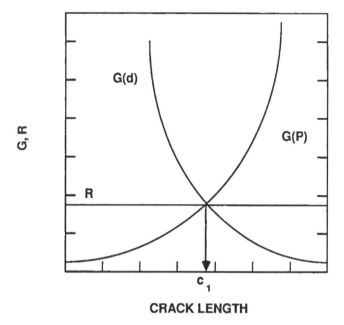

CRACK LENGTH

FIGURE 6.   The crack extension force (G) and the crack resistance force (R) for two different types of loading (fixed d and constant P) as a function of crack length for the DCB loading geometry depicted in Figure 5. A crack of length, c_1, will propagate unstably for constant P, but will arrest for fixed d.

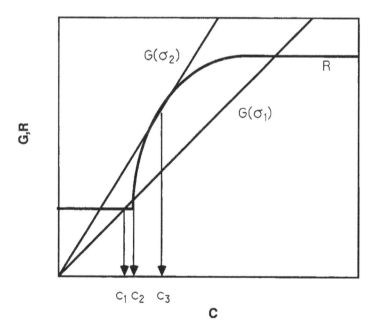

FIGURE 7.   The crack extension force (G) for two different applied stresses and the crack resistance force (R) as a function of crack length for the loading geometry depicted in Figure 1. For this material, R is assumed not to be constant with crack length, but undergoes an increase with crack length. This leads to regions of unstable (c_1 to c_2) and stable crack growth (c_2 to c_3) in a uniform tensile field.

R at c_2. The crack then undergoes a period of stable crack growth as G is increased until, for a value $G(\sigma_2)$ at a crack length c_3, unstable fracture occurs because G = R *and* dG/dc $\geq$ dR/dc. In order to measure R curves in materials, testing geometries in which G decreases with crack length are very important since this leads to stable crack growth. For example, if we were to use a fixed-grip DCB test, we would be able to find the complete R curve, whereas in the tension test, only a portion of the R curve would be measured, such as the region from c_2 to c_3 in Figure 7. R curves have been shown to occur in some ceramics and are expected theoretically, but their exact form is a topic of current research.

The formalism of describing fracture, as outlined in this section, assumes that the materials are essentially linearly elastic to failure and has been very successful in describing brittle failure of many materials. For materials that can undergo extensive inelastic deformation, other approaches, such as the J integral or crack opening displacements, are needed, but this will not be covered in this text. The fracture of ceramics, including ZTC, can usually be described by the linear elastic fracture mechanics (LEFM) parameters G, K, and R, but it is becoming clear that extensive inelastic deformation is possible in ceramics, especially ZTC, and there may be a need in the future to introduce a more sophisticated fracture mechanics formalism.

III. TRANSFORMATION TOUGHENING

As we have seen, materials containing t-ZrO_2 were found to have toughness values greater than many other ceramics. Moreover, it was shown using a variety of techniques, including transmission electron microscopy,[9,10] X-ray diffraction,[11] and Raman microprobe analysis[12] that the fracture surfaces had been transformed to some degree to m-ZrO_2. These observations imply that the stresses associated with the propagating crack had induced the phase transformation and led to an increase in toughness. This raised a variety of questions as to why such a transformation should increase toughness and about which chemical and physical parameters control the toughening increment. The analyses that attempted to answer these questions will be described in this section and they give compelling arguments concerning the way toughness depends on the transformation strains, the size and shape of the transformation zone, and how this toughening develops as the crack propagates. The approach we will take will be to consider the effects that the transformation has on the stress field around a crack tip and how this can be used to predict toughening. We will then consider thermodynamic approaches that have been put forward and show that these predict the same toughening as the stress-based analyses. Equivalence between the energy and stress approaches in describing a toughening mechanism must exist, and although it has only been shown in a few cases, it gives credence to the idea that there is a sound, fundamental scientific basis for understanding the transformation-toughening mechanism.

A. Stress Intensity Approach

In Section II, we indicated that the stresses near a crack tip (Equation 6) are simply determined by the *applied* stress intensity factor, i.e., K_I^A. If a phase transformation involving shear and dilational strains occurs in this region, the near-tip stresses must be changed. Near the tip, the transformation strain is saturated and if the material is linear elastic, the stresses can be described by a *local* stress intensity factor K_I^L. If the stresses are reduced by the transformation ($K_I^L < K_I^A$), the transformation has *shielded* the crack tip from the applied loads. This situation is shown in Figure 8, where near the crack tip the stresses depend on K_I^L, i.e., $\sigma_{ij} = K_I^L f_{ij}(\Theta)/[2\pi r]^{1/2}$, and outside the zone, they depend on K_I^A, as in Equation 6. If we define the change in stress-intensity factor $\Delta K = K_I^L - K_I^A$, shielding occurs if $\Delta K < 0$. We expect crack growth to occur when $K_I^L \geq K_{IC}^0$, the fracture toughness of the fully transformed material. The measured toughness, however, is that determined by a critical

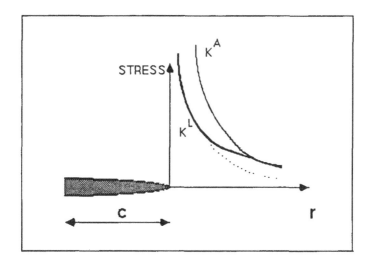

FIGURE 8. In crack-tip shielding, the stresses near a crack tip are reduced locally, so their magnitude is described by a local stress-intensity factor (K_I^L) rather than the applied value (K_I^A).

value of the *applied* stress intensity factor, i.e., $K_I^A = K_{IC}^A$. Therefore, $K_{IC}^A = K_{IC}^0 + \Delta K_{IC}$, where ΔK_{IC} is the value of $-\Delta K$ determined at the critical condition. To predict the increase in fracture toughness due to the transformation compared to that of a fully transformed material, we need, therefore, to evaluate ΔK_{IC}.

The most convenient approach to calculating ΔK_{IC} is to consider a purely dilational (volume changes, but not shape) phase transformation, in which the volumetric strain associated with the unconstrained transformation is e^T. For transformation toughening, this calculation is expected to be reasonable if the m-ZrO_2 particles are extensively and uniformly twinned, thereby eliminating long-range shear stresses. This appears to be the case in several partially stabilized zirconia (PSZ) systems.[13] The value of ΔK_{IC} is dependent on the zone shape and we will consider examples of the stress intensity calculations based on a variety of zone shapes.

1. Frontal Zone

Assume we have a crack in a material that contains only untransformed particles. If we now apply a stress to this material, we can induce the phase transformation. The highest stresses according to the elastic solutions for a sharp crack will be in the region near the crack tip. For a critical stress state (σ_{ij}^c), spontaneous transformation of the particle is expected to occur. Let us assume that transformation occurs once the mean stress (σ_m) acting on a particle reaches a critical value (σ_m^c) to be consistent with the strains in a dilatant transformation. The mean (hydrostatic) stress is defined as one third the sum of the normal stresses. If one assumes the transformation does not alter the stresses outside the zone, one can take the elastic solution for the crack tip stresses and derive an expression for the size of the transformation zone (r_T), which is given by[14]

$$r_T = [2(1 + \upsilon)^2/(9\pi)][K_I/\sigma_m^c]^2 \cos^2(\Theta/2) \qquad (8)$$

The shape of the frontal transformation zone defined by Equation 8 is shown in Figure 9. Once a zone shape is known or assumed, the next step in the calculation is the determination of ΔK. For this discussion, we have made a specific assumption about the zone shape, but as we shall see later, zone shape can influence the predicted value of ΔK. In order to calculate

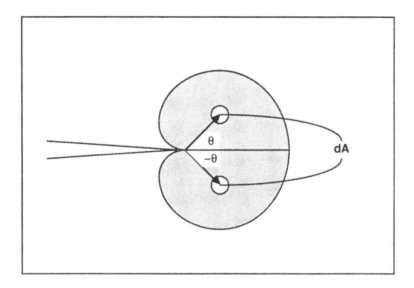

FIGURE 9.   The symmetric cylindrical regions (area dA), that undergo a dilatant transformation within the frontal process zone at angles $\pm \Theta$ to the crack tip.

ΔK, we consider two cylindrical regions of cross-sectional area dA, located at $(r, \pm \Theta)$ with respect to the crack tip, as shown in Figure 9. These two regions, if unconstrained by the matrix, are assumed to undergo a dilation Ve^T, where V is the volume fraction of particles that transform in the zone near the crack tip. That is, cylindrical regions are assumed to contain a volume fraction V of arbitrarily shaped particles that transform and dilate the cylinder. Assuming this is true everywhere in the zone, and that the elastic properties of the transformed material are the same as the untransformed, it is then possible to determine ΔK since $\Delta K = \int\int dK$ over the zone area, where dK is the stress intensity factor produced by the dilating region. The value of dK has been shown to be given by[15]

$$dK = EM \, dA/[(1 - \upsilon^2)(8\pi r^3)^{1/2}] \tag{9}$$

where M is a term that depends on Θ and the stress-free transformation strains. For a purely dilational transformation $M = 2Ve^T(1 + \upsilon)\cos(3\Theta/2)/3$. After performing the double integral, one finds $\Delta K = 0$.[14,16] For a frontal zone of this shape, there is, therefore, *no toughening*. It was shown that this general conclusion is a natural consequence of the path independence of the J integral for such a frontal zone.[17] The conclusion that $\Delta K = 0$ would also be true for partially transformed zones provided contours of constant V follow those of constant σ_m.

It is instructive to consider some of the consequences of Equation 9. For particles that transform in the region $\Theta \leq 60°$, the value of $\Delta K > 0$. This means that transformed particles in this region exert tensile stresses making crack propagation easier. At angles greater than this, $\Delta K < 0$, so that in this region the particles tend to close the crack, increasing toughness. This effect is illustrated in Figure 10. For a dilatant region A directly in front of the crack ($\Theta = 0°$), the surrounding area will be in tangential tension, increasing the crack tip stresses in the y direction. The dilatant regions B ($\Theta = \pm 90°$), however, give rise to radial compression and, thus, the area between the regions will be compressed in the y direction, opposing the tension from A. The variation of the net ΔK is shown schematically in Figure 11 to illustrate the effect of the angle of the dilatant region.[18]

The above calculation coupled the assumptions that both the nucleation and transformation

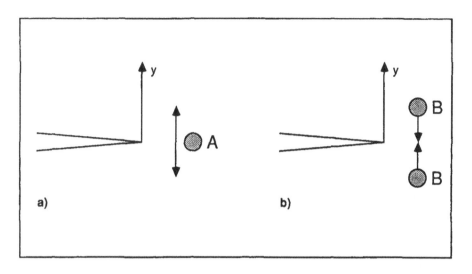

FIGURE 10. The different effect of the dilating cylindrical regions equidistant from a crack tip. In (a) Region A gives rise to tensile stresses in the y direction at a point, while (b) Regions B, at a different angle, give rise to compressive stresses at the same point.

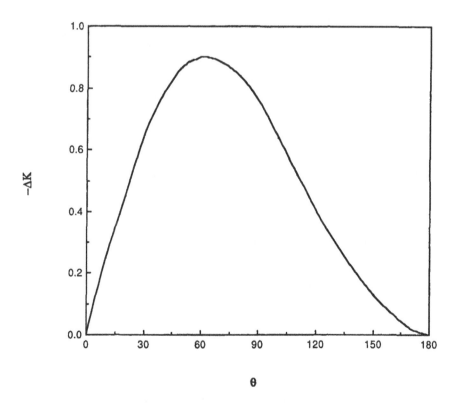

FIGURE 11. The change in the stress-intensity factor at a crack tip as a function of the angle of the dilating region. For small angles ($<60°$) the stress intensity factor increases, but this is offset by the regions at the higher angles, such that overall, the net effect of the frontal zone is zero. (After Marshall, D. B., Drory, M. D., and Evans, A. G., *Fracture Mechanics of Ceramics*, Vol. 5, Bradt, R. C., Evans, A. G., Lange, F. F., and Hasselman, D. P. H., Eds., Plenum Press, New York, 1983, 289.)

strains are dilational. It is, however, possible that these phenomena are not coupled and studies have considered these two assumptions separately. For example, the assumption that the zone shape is determined by a constant σ_m contour is important in obtaining the result that $\Delta K = 0$. If the nucleation were determined by the maximum principal stress in the crack-tip region, the zone shape is flattened in the front compared to a dilation contour[19,20] and this takes part of the region ($\Theta < 60°$) away that gives $\Delta K > 0$, so overall $\Delta K < 0$, and toughening is obtained. The influence of shear stresses on the nucleation condition has been considered in detail[21] and it was shown that toughening is obtained, $\Delta K < 0$, when a shear strain is involved in the nucleation. Similar effects would arise for zones not determined explicitly by the crack-tip field, but by localized stress concentrations. For example, if the transformation occurs by an autocatalytic process, such as shear bands emanating from a crack tip, toughening would be obtained if the shear bands were located at $\Theta > 60°$ or $< -60°$.[19] The other important assumption in the $\Delta K = 0$ calculation was that only the dilational transformation strain was important in calculating ΔK. This appears to be reasonable when the shear strains (shape change) of the transformed particles are relaxed, but in some special cases, it appears that more toughening can occur than that predicted for dilation alone.[20] The result that $\Delta K = 0$ occurs whenever the transformation and nucleation conditions are coupled.[17]

2. Steady-State Zone

The discussion in the last section was concerned with the presence of a frontal transformation zone, but clearly as the crack propagates, it will pass through the transformed zone, and the transformed particles could be left behind the crack tip. There is an important question here, in that particles left in the "wake" of an advancing crack will be under a very low stress compared with the frontal area, so there is the possibility that they could undergo the reverse transformation. This is because the constraint that allowed the particles to remain tetragonal is now returned. Experimentally, it is often found that a significant number of particles remain in the monoclinic phase, but in some cases, reverse transformation is possible. We will assume initially that all the particles that transform in the frontal zone do not undergo the reverse transformation as they enter the wake. The situation as the crack advances into the frontal zone is illustrated in Figure 12 and we can start to see why the transformation leads to toughening. To understand, let us consider the following set of procedures in a "thought" experiment. First, assume that the zone has not yet transformed and we can "cut it out" from the surrounding material. Now, let the zone transform so that it increases in size. We now apply forces to the boundary of the zone until it becomes the same size as it was before the transformation and we can then replace it into its hole. We now let the body forces relax by applying equal and opposite forces. There will, however, still be stresses within the zone because the zone is smaller than it would be if it were unconstrained and these are depicted in Figure 12. As we can see, some of these forces are acting to close the crack, giving rise to a toughening effect. Thus, as the crack enters the compressively stressed zone, it becomes more and more subject to stresses that act to close the crack and promote crack shielding. The type of "cutting" procedure outlined above is that used by Eshelby[22] to solve inclusion problems of arbitrary shape and elasticity and is the basis of the mechanics used in many of the calculations outlined in this section.

Let us consider the magnitude of the toughening for the case in which a crack is completely surrounded by a transformed zone. We will assume again that the mean stress determines the half-height of the zone (h) and that the transformation strain is purely dilational. The value of h can be obtained from Equation 8, when it is realized that the maximum extent of the zone, normal to the crack plane, occurs when $\Theta = 60°$. One finds that

$$h = [3^{1/2}(1 + \upsilon)^2/(12\pi)][K_I/\sigma_m^c]^2 \qquad (10)$$

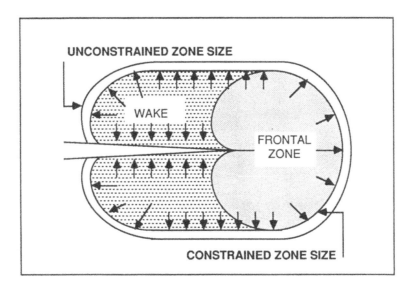

FIGURE 12.   A representation of the stresses that arise in a dilatant transformation zone. The zone must be reduced in size from its unconstrained size to fit in the untransformed material. Once the zone extends behind the crack tip, stresses in the zone arise that are acting in a direction to close the crack.

Performing the double integral for ΔK using Equation 9 over the area of the transformed zone for a semi-infinite crack, one obtains[17]

$$\Delta K_{IC} = 0.2143 E V e^T h^{1/2}/[1 - \upsilon] \tag{11}$$

This result agrees with other studies that used slightly different procedures to obtain ΔK.[14,17,18] We find, therefore, that the magnitude of the toughening from a dilatant transformation depends on the Young's modulus and Poisson's ratio of the untransformed material, the volume fraction of material that transforms, the transformation strain, and the height of the transformed zone. This height can also be expressed in terms of σ_m^c using Equation 10.

In summary, the calculations have shown that for a frontal zone the value of $\Delta K = 0$, but once the zone completely surrounds the crack, in the steady state for a quasi-statically growing crack, there must be an increase in fracture toughness, with the limit for a long crack given by Equation 11. The implication is that the resistance-to-crack growth increases as the crack extends, i.e., "R curve behavior" and we will discuss this in the next section. The cause of the toughening is sometimes stated differently in that the toughening is said to be a result of the transformation zone entering the crack "wake." That is, the toughening is a result of the transformed particles in the crack wake. Later we shall see that even if the transformation does reverse, toughening is still possible.

As with the frontal zone calculation, the assumptions concerning the nucleation conditions (zone profile) and the dominant transformation strains (dilation/shear) can play an important role in the magnitude of the toughening increment. For example, if the transformation was activated in shear bands at 60° to the crack tip, the deleterious transformation in front of the crack tip is removed and the toughening increment is calculated to be[19]

$$\Delta K_{IC} = 0.38 E V e^T h^{1/2}/[1 - \upsilon] \tag{12}$$

The influence of the crack-tip shear stresses on the zone profile has been considered in detail.[21] It was found that slightly higher toughening was predicted when the shear effects

Table 1
RESULTS OF STEADY-STATE TOUGHNESS CALCULATIONS

Net strain coupling	Zone shape	Toughness constant C[a]	Comment
Dilation	Hydrostatic	0.22	Supercritical, no reversal
	Hydrostatic	0.21	Supercritical reverse, max hysteresis
	Shear band profile	0.38	Supercritical, no reversal
Uniaxial dilation	Max principal stress	$0.55(1 - v)$	Supercritical, no reversal
Dilation and relaxed shear strain	Relaxed shear front	0.22	Supercritical, no reversal

[a] $\Delta K_{IC} = CEVe_T h^{1/2}/[1 - v]$

Reprinted with permission from *Acta Metall.*, 34, Evans, A. G. and Cannon, R. M., Toughening of brittle solids by martensitic transformations, Copyright 1986, Pergamon Press plc.

were included. Another case that has been considered is when the transformation is triggered by the maximum principal stress with all the transformation occuring in that direction as well (no adjustment of stress by subsequent twinning within the zone.) The toughening increment is found to be[20]

$$\Delta K_{IC} = 0.55EVe^Th^{1/2} \tag{13}$$

and it follows that special modes of transformation shear within the particles can give more toughening than dilation alone. For cases in which one includes the transformation shear strains, the form of M in Equation 9 becomes more complicated,[15,19,20] but otherwise the approach is similar to that outlined here. The toughening increment has been numerically calculated for the case where the net shear stresses within the particles are relaxed.[20] For oblate spheroidal particles with three orthogonal orientations, typical of PSZ, one orientation produces a larger toughening increment than the dilational case, whereas the other two reduce the shielding effect. For equal proportions of the three orientations, the net effect is a slightly lower toughening increment than the dilational case. The same calculation has been performed for spherical particles and, in this case, the toughening increment slightly exceeds that for the pure dilational case. Thus, the effect of the shear in the particles, unless it is a special mode such as described earlier, appears to be of secondary importance compared to the dilation. The various predicted values of ΔK_{IC} are summarized in Table 1. In this section, we have only considered transforming particles that do **not** undergo the **reverse transformation** as they enter the crack wake and that all particles in the zone transform. Both these effects are not necessarily the case and these will be discussed later as they will also influence the amount of toughening.

3. R-Curve Behavior

It has been shown that for a material with a frontal dilatant zone, the toughening increment is zero, but that for the case of the zone completely surrounding the crack, the toughening approaches a limit given by Equation 11. It follows, therefore, that the crack resistance force, R, increases as the crack propagates into the zone. The change in R as the crack

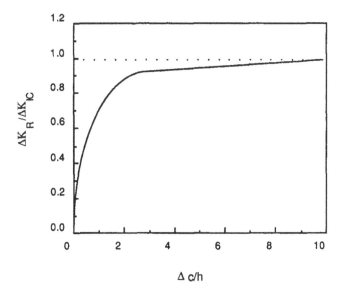

FIGURE 13. The toughening increase (ΔK_R), normalized by its asymptotic value (ΔK_{IC}), that occurs as a crack extends, such that the transformation zone is left behind the crack tip. (After McMeeking, R. and Evans, A. G., *J. Am. Ceram. Soc.*, 65, 242, 1982.)

enters the transformation zone is shown in Figure 13, using a resistance stress intensity factor, (ΔK_R) rather than R. (ΔK_R) is normalized with respect to the toughening limit given in Equation 11. The major increase in (ΔK_R), with increasing crack length Δc, occurs when the transformation zone extends into the crack wake by a distance $\Delta c/h < 5$. The slope of the (ΔK_R) curve has been predicted explicitly for the dilational transformation, assuming h does not change as the crack advances. As can be seen in Figure 13, (ΔK_R) asymptotically approaches the limit given in Equation 11. These R-curve effects are only expected in cases when the wake evolves during crack extension. Presumably, in some cases, the wake may already be present around preexisting cracks unless the material has been subjected to an annealing step. As we saw in Section II, the condition for a crack to become unstable depends on the loading geometry and the particular shape of the R curve, but in general one does not expect R to reach its maximum value, i.e., the actual toughening should be less than the asymptotic limit. As seen in Figure 13, however, the slope of the (ΔK_R) curve is steep and the toughening is expected to be close to the asymptotic limit.

An important aspect of R-curve behavior in a material is that small cracks can grow more easily than large ones and thus, such materials become insensitive to the presence of large, preexisting flaws. It also implies that strength distributions will have less scatter, which is attractive for design purposes. The relationship between R-curve behavior and strength will be discussed further in Section III.E.

B. The Strain-Energy-Release Rate

The thermodynamics of crack advance can be examined by applying a conventional Griffith free-energy approach or by adopting energy-balance integrals. The bases of these two approaches will be discussed separately in this section.

1. A Griffith Approach

As pointed out in Chapter 2, Section VI. B (Equation 6), the total free-energy change per unit volume ($\Delta F_0/V$) of a constrained particle that undergoes a phase transformation is given by

$$(\Delta F_0/V) = \Delta F_{CH} + \Delta U_T \tag{14}$$

where ΔU_T is the sum of the strain energy density and surface energy terms. The term ΔF_{CH} is negative when a particle is below the unconstrained transformation temperature and in some studies a negative sign is used to make this point explicit. In the presence of an external stress an additional interaction term $(-\Delta U_I)$ is required in the energy balance[23]

$$\Delta U_I = \sigma_{ij} e_{ij}^T \tag{15}$$

where σ_{ij} are the stresses applied to the particle and e_{ij}^T are the unconstrained transformation strains. These latter strains must take into account the actual shape change that occurs, i.e., it must account for the actual twin/variant configuration that occurs in the constrained particles. If energy dissipation accompanies the transformation, a further term, ΔU_D, is needed. Taking into account these various terms we have[18]

$$(\Delta F_0/V) = \Delta F_{CH} + \Delta U_T - \Delta U_I + \Delta U_D \tag{16}$$

Spontaneous transformation in the crack tip stress field will occur when $\Delta F_0/V = 0$. If $\Delta U_D = 0$, thermodynamic equilibrium exists at all stages during the transformation and the corresponding set of critical transformation stresses, (σ_{ij}^0) represents a lower bound. More generally, $\Delta U_D > 0$ so that an energy barrier to the transformation exists and the critical set of transformation stresses (σ_{ij}^c), larger than σ_{ij}^0, is needed. For this case the energy dissipated is given by $\Delta U_D = -\Delta\varphi$, where $\Delta\varphi$ is the total change in potential of the transforming particles, i.e.,

$$\Delta\varphi = \Delta F_{CH} + \Delta U_T - \Delta U_I \tag{17}$$

We now return to the energy-balance approach discussed in Section II of this chapter, but incorporating the energy terms associated with the stress-induced phase transformation at the crack tip into the analysis.[18] If we take Equation 3 and incorporate the ideas of Irwin (G and R), the change in free energy (dU) associated with a crack increment dc, for a nontransforming material can be written as

$$dU = G\,dc + R_0\,dc \tag{18}$$

where R_0 is the crack resistance force in the absence of transformation. In this case, crack propagation occurs when the critical crack extension force $G_C^0 = R_0$. If we now incorporate the free-energy terms associated with a stress-induced phase transformation (Equation 16) into the Griffith formalism, we obtain

$$dU = G\,dc + R_0\,dc + d\varphi + dU_D \tag{19}$$

where $d\varphi$ is the change in potential associated with the transformed particles and dU_D is the energy dissipated during the crack increment dc. At Griffith equilibrium, $dU/dc = 0$ and G will reach a critical value G_C and we obtain from Equation 19

$$G_C = R_0 + d\varphi/dc + dU_D/dc \tag{20}$$

The increase in the critical crack extension force due to the transformation $\Delta G_C = (G_C - G_C^0) = (G_C - R_0)$ and, thus, Equation 20 becomes

$$\Delta G_C = d\varphi/dc + dU_D/dc \tag{21}$$

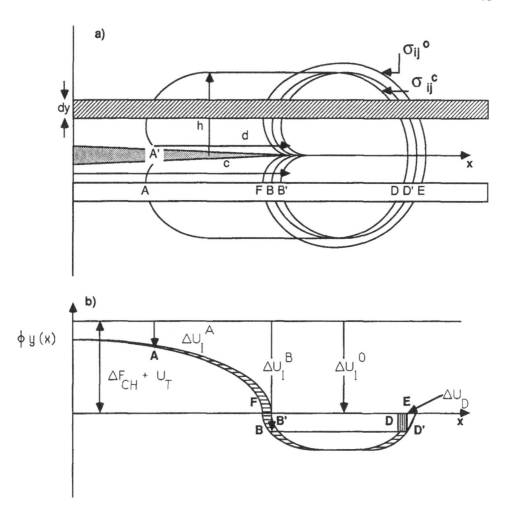

FIGURE 14. (a) The crack/transformation zone configuration used for the energy balance analysis. The crack tip was initially at A′ with no transformation zone, but has extended by an amount d along the crack surface and then subsequently by an amount dc. Before this final increment, the transformation zone shape is shown for two cases of a critical (dilatant) transformation stress, i.e., without (σ_{ij}^0) and with (σ_{ij}^c) a nucleation barrier. (b) The potential energy per unit volume of transformed particles within the shaded strip, width dy, shown in (a) when there is a barrier to the transformation. The location of the boundaries within the shaded strip are shown in the lower strip in (a). The potential curve is shown for two situations, i.e., before and after the final crack increment dc. The change in the potential is represented by the shaded region between the curves. Note that the two lower areas between B and D′ exactly offset each other. (After Marshall, D. B., Drory, M. D., and Evans, A. G., *Fracture Mechanics of Ceramics*, Vol. 5, Bradt, R. C., Evans, A. G., Lange, F. F., and Hasselman, D. P. H., Eds., Plenum Press, New York, 1983, 289.)

The value of R_0 is the toughness in the absence of the transformation, i.e., the material is identical in all respects, except that all the material is transformed before the crack is introduced. Effects involving other toughening mechanisms, such as crack tip interactions, are included in R_0 and the terms involving $d\varphi$ and dU_D are the result of the transformation.

The scheme used for calculating $d\varphi$ is shown in Figure 14a, in which $d\varphi$ is first calculated for the shaded strip of width dy and then is summed over the total width of the transformation zone. Let us consider this figure in detail by considering the changes in φ as we move from right to left in terms of the crack tip stresses. The transformation contour (σ_{ij}^0) represents the outer portion of the transformation zone for the case where there is no barrier to the stress-induced phase transformation, point E on the strip. If there is a barrier to the transformation, a higher stress is required for the transformation and the contour (σ_{ij}^c) denotes the zone boundary in this case. At point D on the strip, the particles then transform. The rear boundary of the (σ_{ij}^c) contour is at point B and this will be the extent of the transformation when there is only a frontal zone with a barrier to the transformation. Point F represents the rear of the frontal zone if there was no barrier to the transformation. When the crack moves forward by an amount dc, these various points, B, D, E, and F will move by an amount dc. Examples are shown in the figure of B moving to B′ and D moving to D′. If the transformation is irreversible, transformed particles are left in the crack wake that extends behind the crack tip. For example, in Figure 14a, the wake extends to a position A_1 along the crack surface or to point A along the strip. Figure 14b shows the variation of dφ for the shaded strip per unit volume of transformed particles as a function of position along the strip, $\varphi_y(x)$. We take the potential of untransformed particles, which is independent of the applied stress, to be zero. At E, which lies on the (σ_{ij}^0) contour, the potentials of untransformed and transformed particles would be equal. Closer to the crack tip, the stress on the particles in the zone increases and, thus, the potential of transformed particles would decrease because of the change in the interaction energy. At point D, the decrease in the potential is sufficient to overcome the energy barrier and the particles transform. Further along the strip, the stresses acting on the particles pass through a maximum so the interaction energy passes through a minimum. At B, the particles experience the same stress (and interaction energy) as when they first transformed. At point F, the rear boundary for a barrierless transformation, the potential will again be zero. In cases where transformed particles are left in the crack wake, the potential continues to rise. In the limit, they will be under a negligible stress and the interaction energy will be zero. For this case, the potential will rise by an amount, given by Equation 15, $\Delta F_{CH} + \Delta U_T$. The potential along the strip $\varphi(y)$ per unit specimen thickness is given by

$$\varphi(y) = V \, dy \int_A^D \varphi_y(x) \, dx \tag{22}$$

where D and A are the front and rear boundaries of the zone and V is the volume fraction of transformed particles. The crack increment displaces the potential in the x direction by an amount dc as the crack extends. The change in potential, dφ(y), is then the shaded areas between the two curves in Figure 14b multiplied by Vdy, i.e.,

$$d\varphi(y) = V \, dy \, dc\{(\Delta U_I^B - \Delta U_I^A) - (\Delta U_I^D - \Delta U_I^0)\} \tag{23}$$

The first difference in the interaction energies comes from the region AB and the second from the region DD′. For the region BD′, the two areas are equal and opposite and the net contribution is zero and, as pointed out earlier, $\Delta U_I^B = \Delta U_I^D$. The second term in Equation 23 is balanced by the energy dissipated by the particles that transform between D and D′ during the crack increment, so

$$dU_D = -\Delta\varphi V \, dy \, dc = (\Delta U_I^D - \Delta U_I^0)V \, dy \, dc \tag{24}$$

The toughening increment, therefore, derives solely from the interaction energy changes in the region AB behind the crack tip and is given by summing Equations 23 and 24 over the width of the zone and then using Equation 21.

$$\Delta G_c = V \int_{-h}^h (\Delta U_I^B - \Delta U_I^A) \, dy \tag{25}$$

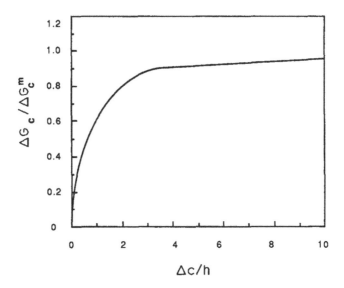

FIGURE 15. The increase in critical crack extension force (ΔG_C) normalized by its maximum value (ΔG_C^m) as the crack extends into a transformation zone. (After Marshall, D. B., Drory, M. D., and Evans, A. G., *Fracture Mechanics of Ceramics*, Vol. 5, Bradt, R. C., Evans, A. G., Lange, F. F., and Hasselman, D. P. H., Eds., Plenum Press, New York, 1983, 289.)

For the case in which there is only a frontal zone, $\Delta U_I^B = \Delta U_I^A$ and it is immediately clear that no toughening would occur. Thus, toughening can be viewed solely in terms of the interaction energy changes behind the frontal zone. We now have to make an assumption about the boundary shape and we again assume that it is determined by the mean stress and that e_{ij}^T is dominated by e^T, as we did in the stress intensity calculation (Section III.A). For these cases, the interaction energy is simply given by $\Delta U_I = \sigma_m e^T$ for a constant mean stress.[23] This allows us to rewrite Equation 25 as

$$\Delta G_C = \Delta G_C^m \left\{ 1 - \int_0^1 (\sigma_m^A / \sigma_m^c) \, d(y/h) \right\} \tag{26}$$

where σ_m^A is the mean stress at the rear of the boundary. The toughening increment, ΔG_C^m, which occurs when the transformation zone extends over the total crack surface is given by

$$\Delta G_C^m = 2Ve^T \sigma_m^c h \tag{27}$$

The variation of ΔG_C as a function of the relative amount of crack extension is shown in Figure 15 and its similarity to Figure 13 and the prediction of R curve behavior is clear. Again the toughening rises rapidly by the time the crack has extended three or four times the zone height. Moreover, if we recall from Section II that $EG = K^2/(1 - v^2)$, we find using Equation 10 that the limit given by Equation 27 is *identical* to that given by the stress intensity calculation, Equation 11.[13,17,18] The calculation outlined here is subject to the same restrictions in terms of zone shape and shear effects as the stress intensity calculation discussed earlier in Section III.A.

2. Energy-Balance Integrals

A more complex, but instructive, approach to determining the value of ΔG_C is the use of

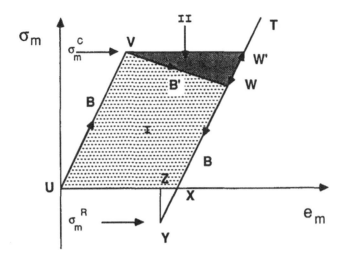

FIGURE 16.   The mean stress-volumetric strain behavior for an element
of constrained material subject to a dilatant phase transformation. The
material initially loads up UV, but at a critical stress (σ_m^c), the transformation occurs down VW. Further loading increases the strain towards T.
Unloading occurs down TY until at Y the material is being subjected to
the residual stress in the far crack wake (σ_m^R). Region I is an area (UVW'X)
for transformation at constant stress. Region II is an area (VWW) associated
with the stress drop during the transformation. (After Budiansky, B.,
Hutchinson, J., and Lambroupolos, J. C., *Int. J. Solids Struct.*, 19, 337,
1983.)

energy-balance integrals[17] and these will be briefly outlined here. It can be shown that the
increment in the crack extension force, due to transformed material being left in the crack
wake, is given by[17]

$$\Delta G = 2 \int_0^h U(y) \, dy \tag{28}$$

where $U(y)$ is the residual strain energy density (residual stress work per unit volume) left
in the crack wake. Comparing Equations 25 and 28, one can see that the $U(y)$ is equivalent
to the interaction energy. The evaluation of $U(y)$ involves knowledge of the stress-strain
behavior of the material in the transformation zone. Figure 16 shows the type of mean stress-
dilational strain behavior expected for a material undergoing a dilatant transformation. This
approach gives one a better sense of what happens to the material during the transformation
in terms of the actual stresses and strains it experiences. It is assumed that the transformed
and untransformed materials possess the same elastic properties. Initially, the loading along
UV is elastic with a slope equal to the bulk modulus B, until at some critical stress, σ_m^c,
the transformation occurs at point V. The transformation, however, does not necessarily
occur at a constant stress, from V to W', and some unloading occurs.[17] This unloading
occurs because even in the absence of an applied stress, such as in the remote crack wake,
the transformed material would be under a residual compressive stress. The slope of the
unloading curve (B') along VW depends on the details of the transformation process and
three types of behavior have been identified.[17] In cases where there is a range of critical
transformation stresses (e.g., as a result of nonuniform particle sizes), the transformation is
termed "subcritical." When there is a barrier to the transformation ($\Delta U_D \neq 0$), the transformation is termed "supercritical". The boundary between these two types of behavior

occurs when there is no barrier to the transformation **and** when all particles transform in the zone. This is termed the "critical" transformation and in this case[17]

$$B' = -2E/(3[1 + \upsilon]) \tag{29}$$

For a subcritical transformation, B' is greater than that given in Equation 29 and for the supercritical transformation, it is less.

Once the transformation has occurred, further loading allows the stress to increase from W towards T and the stress-strain curve has the slope B because the elastic properties of the transformed and untransformed materials are assumed to be the same. If the material is now unloaded, and there is no reverse transformation, the material unloads down the post-transformation slope. For the transformed material under zero stress, the strain at this point (X) will be Ve^T. As the transformed material enters the wake, the zone now contains residual compression and the unloading continues down the elastic curve until one reaches the value of the mean residual stress in the remote wake ($\sigma_m^R = -2EVe^T/\{9[1 - \upsilon]\}$)[17] at point Y. In order to evaluate U(y), the procedure is to determine the area under the stress-strain curve in Figure 16 for a hysteresis loop for the material as it transforms and enters the wake, i.e., the path UVWXY, which includes the energy stored in the wake. This can be written down as the sum of two areas, denoted I and II in the figure, plus the contribution from the wake. Region I (UVW'X) is the area assuming the transformation takes place at constant stress and as drawn; Region II(VWW') must be subtracted from this area to account for the unloading and then a third term is added to account for the energy in the wake, i.e.,

$$U(y) = \sigma_m^c Ve^T + B'(Ve^T)^2/(2[1 - B'/B]) + E(Ve^T)^2/(9[1 - \upsilon]) \tag{30}$$

The subtraction of Region II is taken into account by the negative value of B'. The derivation of the third term is complex because the stress state in the wake contains deviatoric (non-hydrostatic) terms that are not represented in Figure 16. It turns out, however, that the third term in Equation 30 is equivalent to the area UXY.* Using the relationship between B' and E mentioned earlier for a critical transformation (Equation 29), one finds the second two terms, Regional II and III, are equal and opposite and Equation 28 becomes

$$\Delta G_C = 2\sigma_m^c Ve^T h \tag{31}$$

and this is **identical** to Equation 27. It has been shown that Equation 31 also holds for a supercritical transformation provided Equations 28 and 30 are evaluated using Equation 29.[17].

C. Subcritical Transformations and Reversibility

In the toughening calculations outlined above, we have assumed throughout that **all regions** within the zone that are capable of transformation, do so **irreversibly**. Both of these assumptions are questionable and we should examine their consequences. For example, we have already established in Chapter 2 that there is a critical particle size for the transformation and in some materials we expect the particles to vary in size. Thus, we would expect that there will be particles too small to transform within the zone and that as one moves away from the crack tip, the number of particles that have transformed will decrease in these materials (subcritical transformation). If one examines the degree of transformation near a crack with transmission electron microscopy, for example, one finds that only in PSZ systems do all particles transform within the zone.

---

* The third term can be shown to be $3(1-\upsilon)/[2(1-2\upsilon)]$ times the area XYZ and this is equivalent to the area UXY.[58]

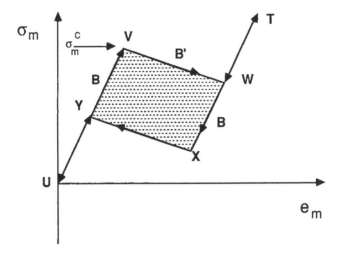

FIGURE 17. The mean stress-volumetric strain behavior for an element of constrained material subject to a dilatant phase transformation for the case where the transformation reverses. In this case, the transformation reverses along XY and the shaded area is the one required to determine U(y).

The use of energy-balance integrals allows the effect of subcritical transformations to be calculated.[17] For a subcritical transformation, the unloading slope (B' in Figure 16) is greater than that in Equation 29, but in addition, the dilational transformation strain will now be less than Ve^T. The actual unloading slope for these calculations depends on the nucleation, growth, and twinning characteristics that occur during the transformation. Some typical situations have been numerically evaluated and these show, as expected, that the maximum toughening increment is less that that obtained for critical and supercritical transformations.[17] This procedure is complex, but fortunately, if one knows the distribution of the transformed particles f(y) as a function of the distance (y) from the crack plane, one can determine the expected increase in toughness. For example, in the case of the dilational transformation one finds[14]

$$\Delta K_{IC} = 0.22 E V e^T h^{1/2} \int_0^h y^{1/2}(df/dy)dy/[1 - \upsilon] \tag{32}$$

This is useful because one can determine f(y) experimentally with techniques such as x-ray diffraction and use this information to determine the toughening increment expected.

It has been shown recently that the reverse transformation (m- to t-ZrO_2) occurs in ZTC under some circumstances.[24] In the discussion of the effect of a frontal zone (Section III.A.1), it is clear that if this zone simply moved with the crack, leaving no wake, the toughening increment would be zero. If, however, the reverse transformation occurs at some point after the material has entered the wake, then toughening still occurs.[18,25] Experimentally, this would lead to difficulties in associating toughening effects with the transformation, since examination of the fracture surfaces would show less transformation than had actually occurred. This would necessitate the use of **in situ** techniques to detect the transformation (during stressing). For example, if we consider the stress-strain curve in Figure 17, it is clear that as one unloads from T, as long as the reverse transformation back to the line UV occurred at a point below W, there would still be a positive hysteresis loop, even though the material returned to point U. Indeed, if the transformation were to reverse at zero applied stress, i.e., when the stress on the transformed material approaches σ_m^R, the toughening would be virtually identical to the irreversible case. This can be seen by referring back to Figure 16, where the reverse transformation would occur along YU. For this case, the third term in Equation 30 is the area XYU, and this does not change the third term in Equation 30 for the critical transformation conditions. In general, if the transformation reverses at an *applied* mean stress σ_m^0, the toughening increment for a supercritical dilatant transformation is simply given by[19]

$$\Delta K_{IC} = 0.22 E V e^T h^{1/2} (1 - [\sigma_m^0/\sigma_m^C])/[1 - \upsilon] \tag{33}$$

For systems in which a reverse transformation takes place, one can obtain toughening but care must be taken in obtaining experimental measurements of the zone size and shape and in the interpretation of the results.

D. Trends in Fracture Toughness

In the various calculations to determine the influence of a stress-induced phase transformation on fracture toughness, we have seen that the amount of toughening depends on the parameters E, V, e^T, and h. On a simple basis, we might expect that we would wish to maximize these four parameters to obtain the maximum toughening, but as we shall see, the situation is rather more complex. On the whole we have seen for ZrO_2 systems that the dilational component e^T of the transformation strains appears to dominate unless one can invoke specialized shear modes. It appears, therefore, that we currently do not have much control over the transformation strains and for the zirconia transformation $e^T \sim 0.04$ to 0.06. We have also determined that an irreversible transformation gives maximum toughening, but that if the transformation does reverse, it is beneficial to do so at as low an applied stress as possible. Finally, we have indicated that the shape of the transformation zone can play an important role. For example, if the nucleation of the transformation can be restrained in the region ahead of a crack tip ($\Theta < 60°$), additional toughening is available. Let us consider in some detail factors that could influence the other factors V, E, and h.

In terms of V, we would like not only to maximize the amount of transformable ZrO_2, but also to make sure it does not undergo a subcritical transformation. This is usually accomplished by making the ZrO_2 particle size distribution as uniform as possible. One could increase E for a material by adding a stiffer second phase, but this could reduce the amount of transformable material in the zone. In addition, a stiffer second phase increases the amount of elastic constraint that resists the transformation and could act to reduce h, by making the transformation more difficult to induce. Moreover, as shown recently,[26] it would reduce the dilatant strain that acts to close the crack. The increased stiffness of the surrounding material increases the residual strain in the transforming particle and reduces it in the matrix. For example, at low volume fractions of the second phase addition, it was shown that Equation 11 must be multiplied by an additional factor F, which is given by

$$F = (B_p + B_m)/[2B_m] \tag{34}$$

where B is the bulk modulus and m and p refer to the matrix and particle. Thus, for a material in which $B_m/B_p = 2$, the toughening increment is reduced by a factor of 0.75.

It is generally accepted that h is the most critical parameter in determining the toughening. However, we have not indicated which parameters will control h, except for indicating that it will increase as the critical transformation stress (σ_{ij}^c) decreases. It is expected that h will be dependent on the nucleation conditions, but as yet, it is not clear how these conditions can be manipulated. It has, however, been shown that end-point thermodynamic calculations[23,27] will at least give some feeling for the observed trends. These calculations indicate that the

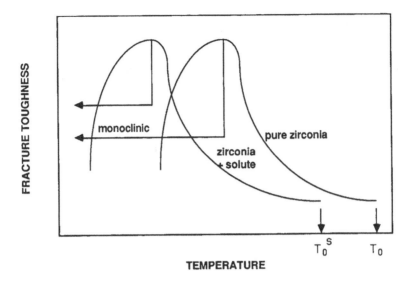

FIGURE 18.   Schematic of the effect of temperature changes and solute addition on the fracture toughness of a transformation-toughened material. The toughness increases with decreasing temperature until the t-ZrO_2 can no longer be retained.

ease of transformation will increase as the amount of supercooling increases. This means the lower the temperature, relative to the temperature at which the transformation would proceed if there was no constraint, the lower σ_{ij}^c and the larger the value of h. Thus, we expect that the fracture toughness of a transformation-toughened material will decrease as the temperature increases. This behavior is shown schematically in Figure 18. At the unconstrained transformation temperature for pure ZrO_2 (T_0), the amount of toughening will be neglibible, but as the temperature decreases, the toughening increases. There will, however, be a limit to this effect; if the temperature is too low, the material will transform in the absence of stress. That is, as the material is cooled down after fabrication, some or all of the t-ZrO_2 particles will transform spontaneously. The amount of supercooling can also be controlled by the presence of solute in the ZrO_2 since this reduces the unconstrained transformation temperature to T_0^s. Thus, at a given temperature, the amount of supercooling will be decreased as one adds the solute and this will decrease the amount of toughening. The benefit of this approach is that the solute will allow the t-ZrO_2 phase to be retained to lower temperatures.

The final item that is important in controlling h is the size of the t-ZrO_2 particles. It has been suggested, as we saw in Chapter 2, that end-point thermodynamic calculations[27,28] can offer an explanation of the size effect. This has been contended by some workers[29,30] and they have indicated that the size effect is related to nucleation. In these cases, the nucleation conditions are related to residual stresses at grain corners or to the statistics of nuclei distribution.[19,30] Whatever the case, the trend in toughness as a function of particle size is reasonably clear once it is accepted that there is a critical size. The expected behavior is shown in Figure 19. At small particle sizes compared to the critical size (d_c), a very high critical stress is needed to overcome the constraint of the surrounding material and thus h and the toughening is small. As the particle size increases, the stress-induced transformation becomes easier and the toughness increases. The maximum toughness would occur if all the particles had a size just below d_c. For larger particle sizes, the transformation has been, to some degree, spontaneous and the amount of t-ZrO_2 available for a stress-induced transformation decreases.

As shown in Figure 19, increasing the amount of t-ZrO_2 will increase the amount of

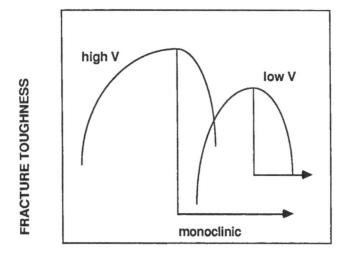

FRACTURE TOUGHNESS

PARTICLE SIZE

FIGURE 19.   Effect of particle size and volume fraction of zirconia on fracture toughness.

toughening that is available, but it decreases the critical particle size. This effect could be a result of interaction effects between particles making the transformation easier (autocatalytic nucleation), or in some systems in which the ZrO_2 is the more compliant phase, it could result from the decrease in the elastic moduli of the material as the ZrO_2 is added.[28,31] A uniform size distribution of t-ZrO_2 particles does have one drawback in that the toughness would tend to be very temperature sensitive. For example, one would have a high toughness at one temperature, but reducing the temperature could lead to a situation in which none of the particles are large enough to transform. The implication in these cases is that h could be rather temperature sensitive. It has also been suggested that particle shape could play a role in nucleating the transformation.[29] In these cases, angular particles are preferred as the stress concentrations associated with corners can aid in nucleation.

Up to this point we have assumed that increasing h will be beneficial in terms of toughness, but it must be realized that there is a *limit* to h above which the *toughness must decrease*. For example, if one reduces the critical transformation stress to such a value that the stress-induced phase transformation occurred throughout the specimen, the transformation zone would no longer be constrained and the crack closure forces would be reduced. The overall trend is shown schematically in Figure 20. The *optimum toughness* occurs when h is large, but small relative to the specimen size.

A key question in the derivation of the various equations for ΔK_{IC} is whether they agree with experimental data and observations. This problem has been discussed recently in detail for PSZ, zirconia toughened alumina (ZTA), and tetragonal zirconia polycrystal (TZP).[19] For PSZ materials, it was concluded that most of the toughening (60 to 80%) could be explained by a dilatant transformation in which the zone shape is determined by the shear stresses and the formation of shear bands (Equation 12). The remaining toughening was associated with microcracking or partial reverse transformation. For ZTA, the data could be accounted for by a subcritical transformation that involves an uniaxial transformation strain, (Equation 13 modified to account for subcritical transformation), but that evidence of this type of transformation was not clear. In some ZTA systems, microcracking appears to play an important role and may be related to the amount of toughening. Microcrack toughening will be discussed briefly at the end of this chapter. For the TZP systems, the

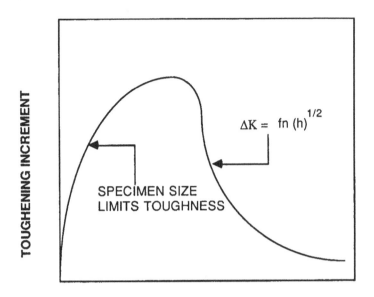

$$\Delta K = fn\ (h)^{1/2}$$

CRITICAL TRANSFORMATION STRESS

FIGURE 20.   Schematic illustration showing effect of critical transformation stress on toughening increment. If transformation stress becomes too low, it will envelop the whole specimen and toughening will be reduced. (After Evans, A. G. and Cannon, R. M., *Acta Metall.*, 34, 761, 1986.)

toughening was associated primarily with the subcritical transformation that involves a uniaxial strain, but in this case Equation 13 underestimated the increase in toughness and it was concluded that the poor agreement was a result of transformation reversal. Finally, it should be noted that both PSZ and ZTA can have appreciable fracture toughness even if the ZrO$_2$ is primarily in the monoclinic form. In these cases, toughening is associated with microcracking and crack deflection.[19] Another useful aspect of microcracked materials is that they often demonstrate good thermal shock resistance, as discussed in Chapter 1.

E. Strength and Toughness

In the first section of this chapter we discussed the relationship between fracture toughness and strength for brittle materials, i.e., $K_{IC} = \sigma_f Y(c)^{1/2}$. From this relationship, it follows that for a constant crack size, increases in K_{IC} will result in an increase in strength. It has, however, been found that materials with the highest fracture-toughness values very rarely possess the highest strengths. In some systems, the complete opposite appears to be the case. A tentative framework[19,32] has been put forward to explain this observation and it invokes an analogy with the mechanical behavior of steels in the ductile-to-brittle transition range.[33]

In our previous discussions in this chapter, we have seen that the toughness should increase as the critical transformation stress (σ_{ij}^c) decreases. Consequently, when brittle failure initiates from preexisting flaws, the strength should increase as σ_{ij}^c decreases. In some cases these preexisting flaws could be associated with transformed ZrO$_2$ particles and may involve the formation of microcracks. At low values of σ_{ij}^c, however, another mechanism can intervene. Specifically, recent evidence suggests that nonlinear deformation, manifested as shear bands, can precede fracture in Mg-PSZ.[34,35] In steels, this behavior leads to microcrack formation as a result of the inhomogeneity of slip in the ductile-to-brittle transition range. Further reduction in the yield stress allows slip to become homogeneous with the onset of ductility,

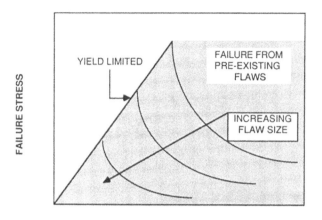

CRITICAL TRANSFORMATION STRESS

FIGURE 21. Schematic illustration showing effect of flaw size and "yield" on the strength of ZTC. (After Evans, A. G. and Cannon, R. M., *Acta Metall.*, 34, 761, 1986.)

which then prevents the microcrack formation. One would expect, in materials such as ZrO_2, that the formation of microcracks at slip band intersections would be more likely since there are not enough slip systems available to allow complete ductility. *In situ* observations of an Mg-PSZ under stress have shown that microcracks do indeed form under some circumstances and they do not become unstable as a result of R-curve behavior.[36] In these circumstances, one would expect that the strength of the material would be "yield" limited, and for a transformation-toughened material, the strength would be simply given by σ_{ij}^c. The overall effect is shown in Figure 21 in which at a constant preexisting flaw size, σ_f increases (σ_{ij}^c decreases), but reaches a maximum when $\sigma_f = \sigma_{ij}^c$. Decreasing flaw size is still useful in that it still allows a higher peak strength. Thus, optimum strength depends on a combination of small preexisting flaw size and an optimum value of σ_{ij}^c. For ZrO_2 materials, it is unlikely that σ_{ij}^c could be reduced sufficiently to allow the onset of complete ductility. An alternative type of diagram has been suggested to describe the type of behavior shown in Figure 21. This diagram as shown in Figure 22 plots strength as a function of fracture toughness. For a given material with a constant inherent flaw size, the strength will increase as the fracture toughness increases due to decreasing σ_{ij}^c, but a peak will be reached when $\sigma_f = \sigma_{ij}^c$ and then the strength will decrease as the fracture toughness further increases. It has been demonstrated that the upper strength boundary is in agreement with experimental data for may different types of ZTC.[34]

The behavior in Figure 22 can also be considered in terms of R-curve behavior.[18,32] In the analysis of transformation toughening, the emphasis was on the maximum toughness increase, but as we indicated in Section II, the condition for crack instability depends on the shape of the R curve (see Figure 7). For example, in Figure 23, the R curves for two materials with different amounts of maximum toughening (different h or V) are shown.[36] This is accomplished by taking the universal curve shown in Figure 15 for different values of ΔG_C^m. If we consider failure in a tensile test, where G is linear with c (Equation 5), the instability condition ($dG/dc = dR/dc$) is given by the tangent of the line through the origin and the R curve. For the R curves shown in Figure 23, we find that the material with the **higher** maximum toughness has the **lower** strength. This conclusion is somewhat tentative since R curves have only been analyzed for long cracks that have no transformation zone before stressing and further research is required in this area. The idea that high-toughness material can possess low strength may seem to be a problem, but as we discussed earlier,

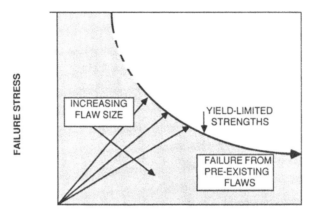

FRACTURE TOUGHNESS

FIGURE 22. Schematic of the relationship between fracture toughness and strength in ZTC showing limit in strength due to R-curve behavior. (After Swain, M. V. and Rose, L. R. F., *J. Am. Ceram. Soc.*, 69, 511, 1986.)

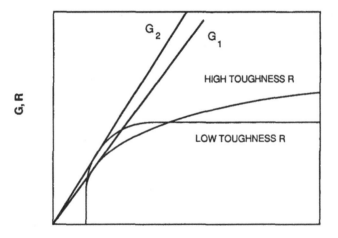

CRACK LENGTH

FIGURE 23. R curve for two transformation-toughened ceramics with different maximum toughness limits. The condition for crack instability given by the point of tangency with a straight line through the origin (uniform tensile field) shows the high toughness material fails at a lower stress (or G). (After Marshall, D. B., *J. Am. Ceram. Soc.*, 69, 173, 1986.)

materials with R-curve behavior will have other benefits, such as tolerance to preexisting flaws and more deterministic strength distributions.

Before we conclude this discussion on strength, it is important to note that strength behavior in ZTC can be complicated by surface compression effects that are associated with trans-formed surfaces. These stresses (up to ~1 GPa) can be introduced in many ways and can lead to strengthening of a material if the strength is dominated by the surface flaw population or if the maximum stress occurs at the surface. The effect of surface stresses will be discussed in more detail in Chapter 6.

IV. OTHER TOUGHENING MECHANISMS

In this chapter, the aim was to emphasize the effects of a stress-induced phase transformation on fracture toughness and strength of ceramics. There are, however, other toughening mechanisms available in ceramics and these will be discussed briefly in the final part of this chapter. In some cases, these mechanisms are thought to occur in ZTC and in other cases, they could be applied to such materials as a source of additional toughening. As pointed out in Section I of this chapter, there are three types of toughening mechanisms that are of current interest in ceramics, viz., crack-tip interactions, crack-tip shielding, and crack bridging, and we will discuss examples of each.

A. Crack-Tip Interactions

The primary aim of this type of mechanism is to place obstacles in the crack path to impede crack motion. These obstacles could be second-phase particles, whiskers, fibers, or possibly, regions that are simply difficult to cleave. One would expect that stress concentrations or residual stresses associated with such obstacles would play a role in this process. There are two different types of consequences that can occur if the crack motion is impeded by an obstacle. In one case, although the crack is pinned by the obstacle, it can bypass the obstacle by "bowing" around either side of it, remaining on virtually the same plane (Figure 24a). In the other case, the crack could attempt to completely avoid the obstacle by "deflecting" out of the crack plane. The deflection of the crack front can be accomplished by the tilting of the crack path or the twisting of the crack front (Figure 24b and c). In a real situation, a combination of bowing and deflection may occur, but let us consider them separately. Both mechanisms have been suggested as playing a role in ZTC, since these are multiphase materials and such obstacles are readily available or can be easily introduced.

1. Crack Bowing

This mechanism has been analyzed theoretically, and it has been shown by two different types of calculations that crack bowing should lead to an increase in fracture toughness.[37,38] There is considerable direct fractographic evidence for such bowing and this has been summarized previously.[39] The analyses of the crack bowing mechanism do have some shortcomings in that they assume the obstacles are impenetrable. One would expect that the strength and toughness of such obstacles would be a key issue, in that either the obstacle could fail before the bowing process is complete or the obstacles are left behind as unbroken ligaments behind the crack tip. In this latter case, crack bowing becomes a precursor to the crack-bridging mechanism. In the former case, it is possible to adjust the calculated toughening to account for obstacle strength.[40] Another important aspect of the crack bowing process that has not been considered is the influence of local stresses around the obstacle on the bowing process.

2. Crack Deflection

If a crack is deflected out of the plane that is normal to the applied tensile stress, the crack will no longer be loaded in a simple Mode I or be subjected to the maximum tensile stress. Two types of deflection have been analyzed: tilting of the crack about an axis parallel to the crack front and twisting about an axis normal to the crack front. The overall deflection process is manifested as roughness of the final fracture surface. The reorientation of the crack plane leads to a reduction of the crack-extension force on the deflected portion.[41] The crack-deflection mechanism was analyzed by evaluating the stress-intensity factors for the three loading modes (Figure 3) on the deflected portion. The various components of K are then combined with a mixed-mode failure criterion and compared to the crack resistance of the matrix that surrounds the obstacle, in order to calculate the increases in fracture toughness.

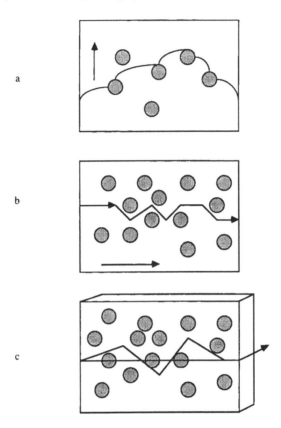

FIGURE 24. A comparison of crack bowing and crack deflection mechanisms. In a, a crack front is shown bowing round a set of obstacles. In b, the crack path is shown for a crack tilting past obstacles. In c, a straight crack front and the twisting it must undergo to avoid a set of obstacles is shown. In each case, the direction of crack propagation is shown with an arrow.

There are several mixed-mode failure criteria and the toughness increase predicted depends on which criterion is utilized. A fracture-mechanics analysis using a fully coupled failure criterion deomonstrated that it is the twist component that contributes most to the fracture toughness.[19,41] For a random array of obstacles, it has been shown that the toughening increment depends on the volume fraction and shape of the particles.[41] The toughening predicted for rod-shaped particles is shown in Figure 25 and shows that rods with large aspect ratios impart maximum toughness. This effect is a result of the increase in twist angle that occurs with increasing aspect ratio.[41] For a given volume fraction of obstacles, shaped as rods, disks, or spheres, it has been shown that rods are the most effective in increasing toughness. As seen in Figure 25, it was also found that most of the toughening develops for volume fractions of obstacles < 0.2. An attractive aspect of the crack deflection mechanism is that it is expected to be independent of temperature and particle size. It should, however, be noted that the fracture mechanics analysis of crack deflection does not account for local stress fields or local changes in the crack resistance force as the crack transverses an obstacle. In the case of localized residual stresses that result from expansion differences, some temperature sensitivity might be expected and these stresses may aid in increasing or decreasing the applied crack extension force. Measurement of fracture toughness values and crack deflection angles in SiC, Si_3N_4, and a glass-ceramic have been shown to be consistent with the predictions of the crack deflection mechanism.[42,43] For ZrO_2 materials, it has been

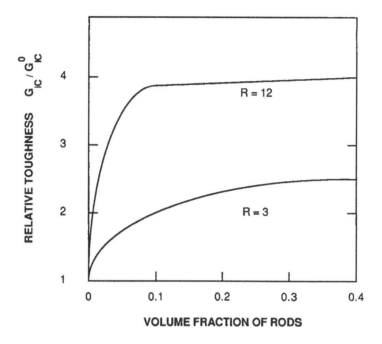

FIGURE 25. The predicted influence of crack deflection on the fracture toughness of a brittle material containing a random array of rod-shaped particles. The influence of increasing the aspect ratio (R) of the rods is included. (After Faber, K. T. and Evans, A. G., *Acta Metall.*, 31, 565, 1983.)

suggested that crack deflection is particularly noticeable around transformed monoclinic particles, especially in PSZ, and this may be an explanation of the relatively high toughness values of the overaged materials.[19]

B. Crack-Tip Shielding

In Section III of this chapter we have already discussed crack-tip shielding, as transformation toughening itself is an example of this type of mechanism. There is, however, another toughening mechanism that involves crack-tip shielding that has been linked to the fracture toughness of ZTC; viz., "microcrack toughening." Ceramics that contain localized residual stresses are known to be capable of microcracking.[44] These residual stresses arise in ceramics as a result of phase transformations, thermal-expansion anisotropy in single-phase materials, and thermal expansion or elastic mismatch in multiphase materials. It is also expected that regions of low toughness, such as grain boundaries, would be attractive sites for such cracks. It has been known for some time that these microcracks can form spontaneously during the fabrication process, provided the grain or particle size is above a critical value. This critical-size effect can be analyzed using fracture mechanics and the results for various situations have been reviewed recently.[44] Ceramics containing microcracks after fabrication have been associated with good thermal shock resistance,[45] but it is expected that such materials would have low strengths as the microcracks are likely failure origins. A more attractive proposition, in terms of fracture toughness, is to fabricate materials in which the particle size is below that required for spontaneous microcracking, but in the range where microcracks could be stress induced. Calculations that consider microcracks forming under an applied tensile stress indicate that particles down to about one half of the critical size might be candidates for the formation of stress-induced microcracks.[44] In terms of fracture, one would expect that the microcracks would form a zone around large cracks, much the same way as a stress-induced phase transformation zone forms. The creation of a microcrack zone around a propagating

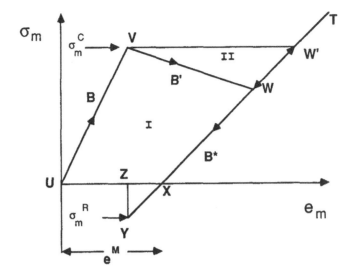

FIGURE 26.   The mean stress-volumetric strain behavior for an element of constrained material subject to a supercritical microcracking. The material initially loads up UV, but at a critical stress (σ_m^c), the microcracking and the material unloads down VW. Further loading increases the strain towards T and the slope of this line is less than that for the nonmicrocracked material. Unloading occurs down TY until at Y the material is being subjected to the residual stress in the far crack wake (σ_m^R). Region I is an area (UVW'X) for microcracking at constant stress. Region II is an area (VWW') associated with the stress drop during the microcracking. (After Evans, A. G. and Faber, K. T., *J. Am. Ceram. Soc.*, 67, 255, 1984.)

crack is expected to reduce the stresses near the crack tip, giving rise to shielding. An alternative view of the toughening phenomena would be to consider the increased amount of fracture surface that must be associated with the formation of such a zone.

The shielding process associated with a microcrack zone is, in some ways, similar to that involved in transformation toughening. In the latter case, we saw in Section III of this chapter that dilation of the process zone was a primary factor in the toughening process. The formation of microcracks is generally associated with residual stress fields and when a crack forms in such a field, it will give rise to a volume increase as the crack opens. Thus, we would expect that if we can predict the volume increase as a function of the microcrack density, the toughening increment could be predicted simply from the transformation-toughening equations. There are, however, two other effects that are important in the shielding process and these are different from any effects discussed in transformation toughening. The first of these effects gives rise to additional shielding and is a result of the decrease in the elastic constants that occurs when a material microcracks. This "modulus" effect will itself reduce the stresses in the zone and, thus, aids in the shielding process.[46] The other effect is based on the realization that the microcracks must degrade the fracture toughness within the zone.[46] That is, the microcracks have "degraded" the material in the zone and fracture surface is already available when the main crack propagates.

The mean stress-volumetric strain behavior of the material as it enters a constrained microcrack zone is shown in Figure 26.[19] We assume the material is initially uncracked with a bulk modulus B and as with the transformation-toughening case, on loading up UV, the material is assumed to microcrack at a critical value of the mean stress (σ_m^c). Again, as with the stress-induced transformation, the material will then unload to W with a slope B'. Further loading to T, however, will involve a different slope B*, which is less than B and this behavior is different than the transformation case. Further unloading will also occur down the slope B* until at zero stress (Point X) we have a dilation e^M. As the applied stress decreases, we can get further unloading down to point Y and the microcracked material is being residually stressed. It is important to note that even in the absence of the dilation, the decrease in slope from B to B* would give rise to a hysteresis loop and, hence, is the reason why the modulus effect gives rise to toughening.

The modulus effect can be analyzed by considering the formation of a microcrack frontal zone. It has been shown using energy-balance integrals for the supercritical case, i.e., when microcracking occurs[46] at all potential sites, that

$$(K_I^L/K_I^A)^2 = (E^*/E)([1 - \upsilon^{*2}]/[1 - \upsilon^2]) \qquad (35)$$

where E and E* are the Young's moduli of nonmicrocracked and microcracked materials, respectively, and υ and υ^* are the associated values of Poisson's ratio. Using Equation 35 we can see the decrease in E and υ will lead to a local reduction in K_I. If we consider the failure criterion, however, we must now take into account the degradation phenomenon associated with the microcracking. This effect must reduce the fracture toughness when compared to that of a nonmicrocracked material (K_{IC}^0). It has been suggested[46] that this degradation is approximately given by

$$K_{IC} = K_{IC}^0(1 - f) \qquad (36)$$

where f is the saturation microcrack density. This latter parameter is actually a dimensionless quantity, Nc^3, where N is the number of (circular) microcracks per unit volume and c is their average radius. The effect of f on elastic constants has been determined previously[47] to be approximately given by

$$E^*/E = 1 - \{16[1 - \upsilon^{*2}][10 - 3\upsilon^*]f/[45(2 - \upsilon^*)]\} \qquad (37)$$

and

$$\upsilon^*/\upsilon = 1 - (16f/9) \qquad (38)$$

Substituting Equations 35, 36, and 37 into the failure criterion $K_I^L = K_{IC}$, one finds[46]

$$K_{IC}^A/K_{IC}^0 = (1 - f)\{1 - [16(1 - \upsilon^{*2})[10 - 3\upsilon^*)f]/45(2 - \upsilon^*)\}^{-1/2} \qquad (39)$$

Using Equation 38, it is clear that the toughening depends only on f. For reasonable values of f (<0.3), one finds that the *toughening increment for a frontal zone* is approximately *zero*. Although this result is the same as the transformation-toughening case, the cause is different in that the degratdation of the material by the microcracks is offset by the shielding modulus effect.

We have just determined that the microcrack-toughening problem turns out to be virtually analogous to the transformation-toughening problem. Thus, if we consider there to be critical value of the *mean* stress required to form microcracks and an associated volumetric strain e^M, the toughening increment for the supercritical case can be found from the results in Section II. For example, we expect the effect on the frontal zone on toughening to be negligible and for the asymptotic case of a fully developed wake that the toughening increment to be the same as Equation 11, i.e.,

$$\Delta K_{IC} = 0.21Ee^M h^{1/2}/(1 - \upsilon) \qquad (40)$$

The value for e^M will depend on the details of the microcracking process. For example, for spheres that microcrack under a residual stress σ_R,[48] it has been shown that

$$e^M = 16(1 - \upsilon^2)\sigma_R V/(3E) \tag{41}$$

where V is the volume fraction of (noninteracting) spheres. Substituting this equation in Equation 40 one obtains[19]

$$\Delta K_{IC} = 1.12V\sigma_R h^{1/2}(1 + \upsilon) \tag{42}$$

For microcracking associated with a phase transformation, in which there is a volume increase, such as in ZrO_2, the microcracks are expected to form radially from the ZrO_2 particles into the matrix. For this situation, it has been suggested[49] for simultaneous transformation and microcracking that

$$e^M = 2e^T(V + [3V/(4\pi)]^{1/3})/3 \tag{43}$$

Substitution in Equation 40 gives

$$\Delta K_{IC} = 0.21Ee^T h^{1/2}(V + 0.6V^{1/3})/[1 - \upsilon] \tag{44}$$

Comparison of Equations 11 and 44 shows that for similar process zone widths, simultaneous microcracking and transformation is a more effective shielding process than transformation alone, but that overall, transformation shielding is the more effective process.

From the above discussion, it is also expected that R-curve behavior should occur in a microcracking material as $\Delta K_{IC} \sim 0$ for a frontal zone, but as the microcracks enter the crack wake, dilation toughening will occur.[46] The relationship between h and the critical microcracking stress is, as yet, poorly understood, but it will involve the flaw populations present in the residual stress fields and the mechanisms by which they nucleate and form the microcracks.

The overall trends in the fracture toughness increment due to microcracking are very similar to those of transformation toughening. For example, as shown in Figure 27, the toughness is expected to increase with particle size up to the critical particle size for spontaneous microcracking (compare Figure 19). For larger particle sizes, the material is already microcracked prior to the application of stress and further toughening will only arise if one can increase the microcrack density f. In a similar way, the width of the particle size distribution will play a role in determining the absolute toughness and one has to deal with subcritical microcracking. The temperature sensitivity could be somewhat different than discussed for the stress-induced phase transformation, depending on the source of the residual stress. If the microcracks are a result of thermal expansion mismatch, then one would expect that increasing temperature would result in a decrease in toughness, as the magnitude of the residual stress will decrease (cf. Equation 42). On the other hand, if the microcracks were formed by a phase transformation, the toughening effect would be relatively temperature insensitive unless the transformation reverses. Increasing the volume fraction of the microcracks is clearly attractive, but again, interaction effects may increase the likelihood of spontaneous transformation and microcrack linking could be a prime source of preexisting flaws in a material. The factors that control the size of the microcrack zone are expected to be similar to those that control spontaneous microcracking and these include the magnitude of the residual stress, the particle size, and the size (and shape) of defects within the residually stressed regions.[44] In a similar way to transformation toughening, one would expect that there will be a limit to the toughening if the microcrack zone size approaches the specimen

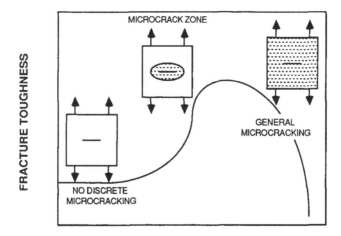

FIGURE 27. Schematic illustration of the effect of particle size on fracture toughness of a microcracking material. For small particles, the critical microcracking stress is too high to form a zone around a crack. As one approaches the critical size, the toughness increases as the zone height increases. At large particle sizes, the material contains microcracks before stressing. (After Evans, A. G. and Faber, K. T., *J. Am. Ceram. Soc.*, 67, 255, 1984.)

size and a "yield limit" to the strength if the critical microcracking stress becomes too small. Nonlinear elastic behavior associated with microcracking has been observed in overaged Ca-PSZ[50,51] and microcrack toughening has been considered to be an important toughening mechanism in ZTA systems.[19]

C. Crack Bridging

In the discussion of crack-tip interactions, we indicated that a crack may bypass an obstacle, leaving it intact. In such cases, the obstacle is left as a ligament behind the crack tip. In other cases, ligaments may be formed by the mechanical interlocking of the grains. These ligaments will make it more difficult to open the crack at a given applied stress and will increase fracture toughness. This mechanism has been shown to be important in frictionally bonded fiber composites,[52,53] in large-grained Al$_2$O$_3$[54,55] and may also be operational in whisker-reinforced ceramics. In PSZ, some crack bridging by the precipitates has been observed (see Chapter 4, Section IV.C). In order to determine the effect of crack bridging on fracture toughness, it is necessary to know the force-displacement relationships for the ligaments. The effect of crack bridging is best understood for frictionally bonded fiber composites and we will briefly discuss the progress in this area.

The addition of continuous carbon or SiC fibers to glasses or ceramics has been studied since about 1970 and it has been found, in some cases, that unidirectional-fiber composites do not undergo catastrophic failure in tensile loading. This "ductile" type of behavior for a material composed of two brittle components is particularly attractive for structural applications. In the optimum materials, the tensile-loading behavior is initially elastic until at a particular stress a crack passes through the matrix. This crack, however, bypasses the fibers and leaves them available for load carrying and they completely bridge the crack (Figure 28). Further loading causes the formation of regularly spaced, bridged, matrix cracks until at the peak load, the fibers fail. The ensuing failure, however, is not catastrophic as the fibers continue to pull out of the matrix.

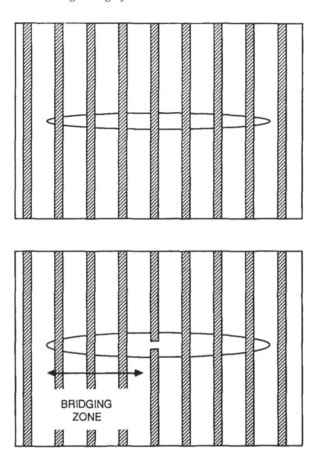

FIGURE 28.   Comparison of a crack fully bridged by frictionally bonded fibers with the case where fibers break during matrix cracking forming a bridging zone behind the moving crack front.

In these materials the final failure is not the result of the propagation of a single crack and, thus, a fracture toughness value cannot be defined. It is, however, possible to use fracture mechanics to describe the conditions at which the first crack passes through the matrix.[56] In this analysis, it was found that the matrix cracking stress (σ_0) is independent of the preexisting flaw size and, thus, a material property. The analysis also identified the microstructural parameters that can be used to control σ_0. Techniques to increase σ_0 are attractive, but if this stress approaches that of the fibers, then fiber failure will accompany matrix cracking and there will be a transition to a brittle type of failure.[57] For these cases, a bridging zone of limited extent is formed behind the crack and moves with the crack. The overall effect of crack bridging on the strength of fiber composites can be shown in normalized plot (Figure 29). The normalization parameters take into account the specific material properties involved in the process. For a fully bridged crack, it was found that above a certain crack size, the strength becomes independent of crack size, as discussed earlier. For partially bridged cracks, the bridging increases strength for a given crack size, but the strength is no longer independent of crack size.

In this chapter, we have discussed the principles underlying fracture of ceramics and have shown the way this framework can be used to analyze the available toughening mechanisms. The importance of these types of analyses is that they confirm that a particular mechanism is of sufficient magnitude to explain the observed toughening effects, but perhaps, more

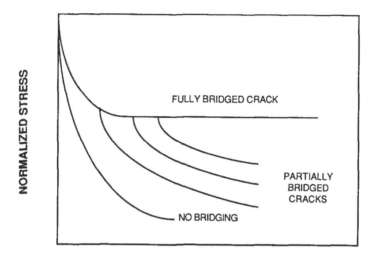

NORMALIZED CRACK LENGTH

FIGURE 29. The strength of a frictionally bonded, unidirectional fiber composite as a function of crack size. For a fully bridged crack, there is a region where strength does not depend on crack size. For cases where the fibers break during matrix cracking, the strength does depend on crack size and the magnitude of the strengthening depends on the size of the bridging zone. (After Marshall, D. B. and Evans, A. G., *Proc. 5th Int. Conf. on Composite Materials,* Harrigan, W. C., Strife, J., and Dhingra, A. K., Eds., Metallurgical Society, Warrendale, Pa., 1985, 557.)

importantly, they clearly identify the parameters that need to be controlled and optimized to produce improved materials. We now turn to the actual mechanical behavior observed in the various types of transformation-toughened systems based on ZrO_2.

REFERENCES

1. **Griffith, A. A.,** The phenomena of rupture and flow in solids, *Philos. Trans. R. Soc. London,* A221, 163, 1920.
2. **Inglis, C. E.,** Stresses in a plate due to the presence of cracks and sharp corners, *Inst. Naval Architects Trans.,* 55, 219, 1913.
3. **Irwin, G. R.,** Fracture, *Handbuch der Physik,* 6, 551, 1958.
4. **Westergaard, H. M.,** Bearing pressures and cracks, *J. Appl. Mech.,* 61, A49, 1939.
5. **Sih, G. C.,** *Handbook of Stress Intensity Factors,* Lehigh University Press, Bethlehem, Pa., 1973.
6. **Tada, H., Paris, P. C., and Irwin, G. R.,** *The Stress Analysis of Cracks Handbook,* Del Research Corporation, St. Louis, Mo., 1973.
7. **Lawn, B. R. and Wilshaw, T. R.,** *Fracture of Brittle Solids,* Cambridge University Press, Cambridge, Mass., 1975.
8. **Green, D. J.,** Mechanical behavior of ceramics, *CerSE 414 Course Notes,* Pennsylvania State University, University Park, 1987.
9. **Porter, D. L. and Heuer, A. H.,** Mechanism of toughening partially stabilized zirconia (PSZ), *J. Am. Ceram. Soc.,* 60, 183, 1977.
10. **Ruhle, M., Strecker, A., Waidelich, D., and Kraus, B.,** *In situ* observations of stress-induced phase transformation in ZrO_2-containing ceramics, in *Science and Technology of Zirconia II,* Advances in Ceramics, Vol. 12, Claussen, N., Ruhle, M., and Heuer, A. H., Eds., American Ceramic Society, Columbus, Ohio, 1984, 256.
11. **Kosmac, T., Wagner, R., and Claussen, N.,** X-ray determination of transformation depths in ceramics containing tetragonal zirconia, *J. Am. Ceram. Soc.,* 64, C72, 1981.

12. **Clarke, D. R. and Adar, F.**, Measurement of the crystallographically transformed zone produced by fracture in ceramics containing tetragonal zirconia, *J. Am. Ceram. Soc.*, 65, 284, 1982.

13. **Evans, A. G.**, Toughening mechanisms in zirconia alloys, in *Science and Technology of Zirconia II*, Advances in Ceramics, Vol. 12, Claussen, N., Ruhle, M., and Heuer, A. H., Eds., American Ceramic Society, Columbus, Ohio, 1984, 193.

14. **McMeeking, R. and Evans, A. G.**, Mechanics of transformation toughening in brittle materials, *J. Am. Ceram. Soc.*, 65, 242, 1982.

15. **Hutchinson, J. W.**, On steady quasi-static crack growth, *Harvard University Report*, DEAP S-8, Division of Applied Sciences, Cambridge, Mass., 1974.

16. **Seyler R. J., Lee, S., and Burns, S. J.**, A thermodynamic approach to fracture toughness in PSZ, in *Science and Technology of Zirconia II*, Advances in Ceramics, Vol. 12, Claussen, N., Ruhle, M., and Heuer, A. H., Eds., American Ceramic Society, Columbus Ohio, 1984, 213.

17. **Budiansky, B., Hutchison, J., and Lambroupolos, J. C.**, Continuum theory of dilatant transformation toughening in ceramics, *Int. J. Solids Struct.*, 19, 337, 1983.

18. **Marshall, D. B., Drory, M. D., and Evans, A. G.**, Transformation toughening in ceramics, in *Fracture Mechanics of Ceramics*, Vol. 5, Bradt, R. C., Evans, A. G., Lange, F. F., and Hasselman, D. P. H., Eds., Plenum Press, New York, 1983, 289.

19. **Evans, A. G. and Cannon, R. M.**, Toughening of brittle solids by martensitic transformations, *Acta Metall.*, 34, 761, 1986.

20. **Lambroupolos, J. C.**, Shear, shape and orientation effects in transformation toughening, *Int. J. Solids Struct.*, 22, 1083, 1986.

21. **Lambroupolos, J. C.**, Effect of nucleation on transformation toughening, *J. Am. Ceram. Soc.*, 69(3), 218, 1986.

22. **Eshelby, J. D.**, The determination of the elastic field of an ellipsoidal inclusion, and related problems, *Proc. R. Soc. London*, 241A, 376, 1957.

23. **Evans, A. G. and Heuer, A. H.**, Review — transformation toughening in ceramics and martensitic transformations in crack-tip stress fields, *J. Am. Ceram. Soc.*, 63, 241, 1980.

24. **Marshall, D. B. and James, M. R.**, Reversible stress-induced martensitic transformation in ZrO_2, *J. Am. Ceram. Soc.*, 69, 215, 1986.

25. **Coyle, T. W. and Cannon, R. M.**, Contributions to toughening from reversible martensitic transformations, abstract, *Am. Ceram. Soc. Bull.*, 60, 377, 1981.

26. **McMeeking, R. M.**, Effective transformation strain in binary elastic composites, *J. Am. Ceram. Soc.*, 69, C301, 1986.

27. **Lange, F. F.**, Transformation toughening: thermodynamic approach to phase retention and toughening, in *Fracture Mechanics of Ceramics*, Vol. 5, Bradt, R. C., Evans, A. G., Lange, F. F., and Hasselman, D. P. H., Eds., Plenum Press, New York, 1983, 255.

28. **Lange, F. F. and Green, D. J.**, Effect of inclusion size on the retention of tetragonal ZrO_2: theory and experiments, in *Science and Technology of Zirconia*, Advances in Ceramics, Vol. 3, Heuer, A. H. and Hobbs, L. W., Eds., American Ceramic Society, Columbus, Ohio, 1981, 217.

29. **Ruhle, M. and Heuer, A. H.**, Phase transformations in ZrO_2-containing ceramics. II. The martensitic reaction in t-ZrO_2, in *Science and Technology of Zirconia II*, Advances in Ceramics, Vol. 12, Claussen, N., Ruhle, M., and Heuer, A. H., Eds., American Cermaic Society, Columbus, Ohio, 1984, 14.

30. **Chen, I.-W. and Chiao, Y.-H.**, Theory and experiment of martensitic nucleation in ZrO_2-containing ceramics and ferrous alloys, *Acta Metall.*, 33, 1827, 1985.

31. **Green, D. J.**, Critical microstructures for microcracking in Al_2O_3-ZrO_2 composites, *J. Am. Ceram. Soc.*, 65, 610, 1982.

32. **Swain, M. V. and Rose, L. R. F.**, Strength limitations of transformation-toughened zirconia alloys, *J. Am. Ceram. Soc.*, 69, 511, 1986.

33. **Hahn, G. T., Averbach, B. L., Owen, W. S., and Cohen, M.**, Initiation of cleavage microcracks in polycrystalline iron and steel, in *Fracture*, Averbach, B. L., Felbeck, D. K., Hahn, G. T., and Thomas, D. A., Eds., John Wiley & Sons, New York, 1959, 91.

34. **Swain, M. V.**, Inelastic deformation of Mg-PSZ and its significance for strength-toughness relationships of zirconia toughened ceramics, *Acta Metall.*, 33, 2083, 1985.

35. **Larsen, D. C. and Adams, J. W.**, Long term stability and properties of zirconia ceramics for heavy duty diesel engines, final contract report, NASA CR-174943, IITRI Technical Report MO 6117, IIT Research Institute, Chicago, 1985.

36. **Marshall, D. B.**, Strength characteristics of transformation-toughened zirconia, *J. Am. Ceram. Soc.*, 69, 173, 1986.

37. **Lange, F. F.**, Interaction of a crack front with a second phase dispersion, *Philos. Mag.*, 22, 983, 1970.

38. **Evans, A. G.**, The strength of brittle materials containing second-phase dispersions, *Philos. Mag.*, 26, 1327, 1972.

39. **Green, D. J.**, Crack-Particle Interactions in Brittle Composites, Ph.D. thesis, McMaster University, Hamilton, Ontario, 1976.
40. **Green, D. J.**, Fracture toughness predictions for crack bowing in brittle particulate composites, *J. Am. Ceram. Soc.*, 66, C4, 1983.
41. **Faber, K. T. and Evans, A. G.**, Crack deflection processes. I. Theory, *Acta Metall.*, 31, 565, 1983.
42. **Faber, K. T. and Evans, A. G.**, Crack deflection processes. II. Experiment, *Acta Metall.*, 31, 577, 1983.
43. **Faber, K. T. and Evans, A. G.**, Intergranular crack deflection processes in silicon carbide, *J. Am. Ceram. Soc.*, 66, C94, 1983.
44. **Green, D. J.**, Microcracking mechanisms in ceramics, in *Fracture Mechanics of Ceramics*, Vol. 5, Bradt, R. C., Evans, A. G., Lange, F. F., and Hasselman, D. P. H., Eds., Plenum Press, New York, 1983, 457.
45. **Hasselman, D. P. H.**, Unified theory of thermal shock resistance of ceramic materials, *J. Am. Ceram. Soc.*, 52, 600, 1969.
46. **Evans, A. G. and Faber, K. T.**, On the crack growth resistance of microcracking brittle materials, *J. Am. Ceram. Soc.*, 67, 255, 1984.
47. **Budiansky, B. and O'Connell, J.**, Elastic moduli of a cracked solid, *Int. J. Solids Struct.*, 12, 81, 1976.
48. **Hutchinson, J. W.**, Crack tip shielding by microcracking in brittle solids, *Acta Metall.*, 35, 1605, 1987.
49. **Faber, K. T.**, Microcracking contributions to the toughness of ZrO_2-based ceramics, in *Science and Technology of Zirconia II*, Advances in Ceramics, Vol. 12, Claussen, N., Ruhle, M., and Heuer, A. H., Eds., American Ceramic Society Columbus, Ohio, 1984, 293.
50. **Green D. J., Embury, J. D., and Nicholson, P. S.**, Fracture toughness of a partially stabilized ZrO_2 in the system CaO-ZrO_2, *J. Am. Ceram. Soc.*, 56, 619, 1973.
51. **Green, D. J. and Nicholson, P. S.**, Microstructural development and fracture toughness of calcia partially stabilized zirconia, in *Fracture Mechanics of Ceramics*, Vol. 2, Bradt, R. C., Hasselman, D. P. H., and Lange, F. F., Eds., Plenum Press, New York, 1973, 541.
52. **Aveston, J., Cooper, G. A., and Kelly, A.**, Single and multiple fracture, in *Properties of Fiber Composites*, IPC Science and Technology Press Ltd., Surrey, England, 1971, 15.
53. **Marshall, D. B. and Evans, A. G.**, Failure mechanisms in ceramic-fiber/ceramic matrix composites, *J. Am. Ceram. Soc.*, 68, 225, 1985.
54. **Steinbrech, R. Khehans, R., and Schaarwachter, W.**, Increase of crack resistance during slow crack growth in Al_2O_3 bend specimens, *J. Mater. Sci.*, 18, 265, 1983.
55. **Swanson, P. L., Fairbanks, C. J., Lawn, B. R., Mai, Y.-W., and Hockey, B. J.**, Crack-interface grain bridging as a fracture resistance mechanism in ceramics, *J. Am. Ceram. Soc.*, 70, 279, 1987.
56. **Marshall, D. B. and Evans, A. G.**, The mechanics of matrix cracking in brittle-matrix fiber composites, *Acta Metall.*, 33, 2013, 1985.
57. **Marshall, D. B. and Evans, A. G.**, Tensile failure of brittle-matrix fiber composites, in *Proc. 5th Int. Conf. on Composite Materials*, Harrigan, W. C., Strife, J., and Dhingra, A. K., Eds., Metallurgical Society, Warrendale, Pa., 1985, 557.
58. **Marshall, D. B.**, Personal communication.

Chapter 4

MICROSTRUCTURE-MECHANICAL BEHAVIOR OF PARTIALLY STABILIZED ZIRCONIA (PSZ) MATERIALS

I. INTRODUCTION

Partially stabilized zirconia (PSZ) is often termed the "conventional" zirconia ceramic, forming the first row in the classification of zirconia-toughened ceramics (ZTC), which was discussed in Chapter 1, see Figure 6. The feature common to all PSZ systems is that zirconia forms the host matrix with the "stabilizer" or alloy additive as the minor component. PSZ materials may be further subdivided into two main groups according to the form and distribution of the metastable tetragonal phase. One group has the tetragonal phase as a precipitated phase confined within a cubic stabilized-zirconia matrix. We shall refer to these as PSZ; they generally contain CaO, MgO, their mixtures or Y_2O_3 as the stabilizing additive, and are considered coarse-grained ceramics with grain sizes in the range of 30 to 60 μm. The second PSZ group has the metastable tetragonal phase occurring as small grains (0.5 to 5 μm) within the fabricated body. The body may be entirely tetragonal or contain grains of a cubic stabilized or a stable tetragonal phase* and are referred to as tetragonal zirconia polycrystals (TZP), containing stabilizing additives such as Y_2O_3 or rare earth oxides (ReO), e.g., CeO_2.

Another class of PSZ materials described in Chapter 1, Figure 6 is that of single crystals. These have microstructures similar to coarse-grained PSZ ceramics but are usually manufactured from the melt. Table 1 briefly summarizes the essential features of the three groups.

In this chapter we shall examine the salient microstructural features and mechanical properties of the various PSZ systems, according to stabilizer types, and also show how these properties are influenced by thermal treatment. Before examining the specific stabilizing systems, it is appropriate to describe the general features of PSZ ceramic fabrication and consider these in terms of the various thermal treatment routes.

II. THERMAL TREATMENTS FOR PSZ

In Chapter 2, Section V, we presented the phase diagrams of the common stabilizers and the requirements for retention of the tetragonal phase to low temperatures. The important feature, obtainable from the phase diagrams, for production of PSZ materials is an accurate knowledge of the phase boundaries and the extent of the phase fields. This knowledge of composition and phase boundary position is essential if a material is to be produced on a commercial basis. Chapter 2, Figure 18b shows a "working" phase diagram of the ZrO_2-CaO system where some of the commercial considerations are mentioned. A further consideration, when choosing a commercial composition, is that the composition and thermal treatment must yield sufficient volume fraction of metastable tetragonal phase after the material is optimally treated (see Sections III and IV). The total volume fraction of tetragonal phase may be determined from a simple lever rule calculation from a phase diagram such as shown in Figure 1a. For example, when a material of composition X% metal oxide, which has been solution treated just inside the c-phase boundary at SST, and is subsequently aged at AT, the equilibrium fraction of t-ZrO_2 will be (BX)/(AB).

---

* The stable tetragonal phase,[1] now referred to as t′, described by Scott[2] as a metastable "nontransformable" high Y_2O_3-content tetragonal zirconia solid solution. The phase is further described in Section VIII.

Table 1
CHARACTERISTICS OF DIFFERENT TYPES OF PSZ
CERAMICS

Coarse grained	Fine grained (TZP)	Single crystals
Matrix: cubic-ZrO_2	Grains: t-ZrO_2	Matrix: c-ZrO_2
Precipitates: t-ZrO_2	Grain size ≤ 1 μm	Precipitates: t-ZrO_2
Grain size: 10—100 μm	Stabilizer Y_2O_3 or	Stabilizer MgO, CaO,
Stabilizer MgO, CaO,	ReO	Y_2O_3, or ReO
Y_2O_3, or ReO		

A. PSZ Fabrication/Precipitation/Coarsening

To illustrate the fabrication process of PSZ ceramics we shall consider the thermal treatment of a hypothetical binary-eutectoid system before discussing some of the individual features of the CaO-, MgO-, CaO/MgO-, and Y_2O_3-PSZ systems.

The starting powders are usually high-purity zirconia and stabilizing oxides of ≤ 1 μm size. The main, detrimental impurity in the starting powders is normally silica. This contaminant causes destabilization of the cubic zirconia matrix by reacting with the stabilizer. At low levels (<0.2%), additives may be used to purge the silica from the sintering zirconia compact[3] (see Section IV).

Figure 1a schematically depicts a "typical" ZrO_2 + low-solute solubility-stabilizer phase diagram (we shall quote actual results for the various systems where pertinent), while Figure 1b depicts that of the higher solubility PSZ materials, in this case the Y_2O_3-ZrO_2 system. Returning now to the low-solute solubility system, the heating cycle of the selected composition (X), which includes the sintering process, will terminate just inside the cubic phase field, and is designated the solid solution temperature (SST). For the Mg-PSZ system the solution/sintering temperature is in the range of 1680 to 1800°C, depending upon the MgO composition.

Experimental observations have shown that during firing most of the sintering occurs between 1150 to 1400°C for Mg-PSZ, as shown in Figure 2a. However it is not until above 1400°C that grain growth occurs, as shown in Figure 2b. The onset of grain growth coincides with the formation of the cubic zirconia phase.[4] A similar behavior has also been observed for Y-PSZ materials, where rapid grain growth occurred during firing in the t + c-ZrO_2 phase field with the c-ZrO_2 grains growing more rapidly.[5] It is thus apparent that the requirement of a high-solid solution temperature results in the coarse-grained nature of the PSZ materials. After firing is complete, the material may be cooled in a number of ways. The prime aim of the cooling cycle is to achieve:

1. A uniform distribution of tetragonal precipitates within the grains.
2. Retention of the majority of the precipitates as the tetragonal phase to room temperature.
3. Prevention of massive proeutectoid zirconia precipitating at the grain boundaries, i.e., the nucleation and growth of t-ZrO_2 at the grain boundaries. This phase may grow to an extent where it will transform to m-ZrO_2 on cooling and so weaken the fabricated body.

The three aims are interdependent and may be achieved by a number of routes.

A uniform distribution of precipitates and the prevention of proeutectoid zirconia are most readily attained by rapid undercooling, below the single phase field boundary, such that a classical homogeneous nucleation and precipitation process becomes operative. If cooling is too slow, a large proeutectoid zirconia grain boundary phase results, due to heterogeneous

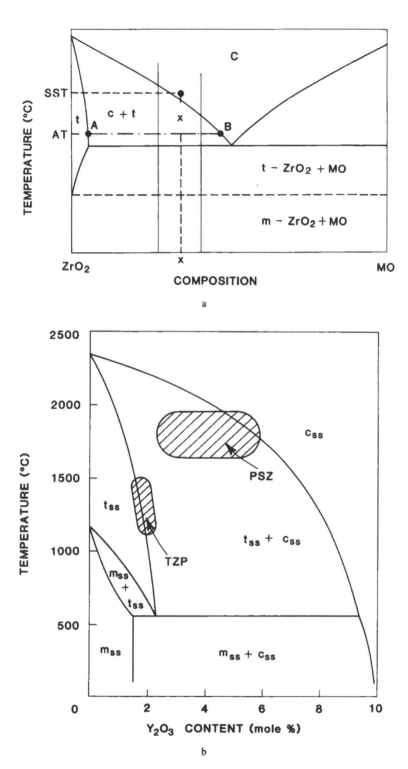

FIGURE 1. Typical phase diagrams for ZrO_2 alloys with low and high solute solubility in the Zr lattice at low temperatures. (a) schematic phase diagram of a ZrO_2-MO (M = Mg, Ca) system, showing the solid solution temperature (SST) and aging temperature (AT) for an alloy of composition X with low solute solubility; (b) partially stabilized zirconia (PSZ) and tetragonal zirconia polycrystal (TZP) compositions with high solute solubility and sintering/solid solution temperatures (shaded regions). In this example we have shown part of the ZrO_2-Y_2O_3 system.

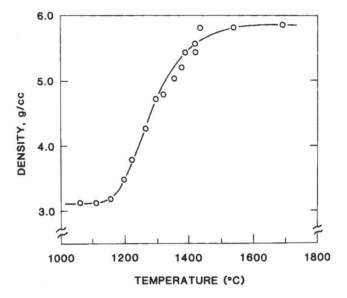

a

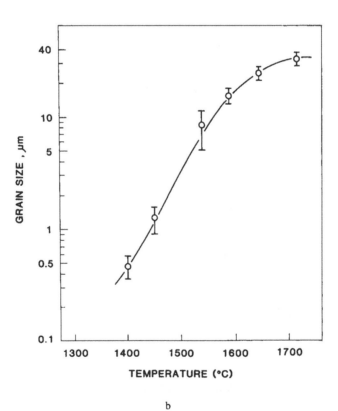

b

FIGURE 2.   Effect of treatment temperature on Mg-PSZ bodies showing
(a) the change in density and (b) grain size.

nucleation. For homogeneous nucleation, the rate of nucleation, N, is strongly dependent on the undercooling, ΔT, through the free energy term, ΔF^*, such that

$$N \ \alpha \ \exp(-\Delta F^*/kT) \quad \text{and} \quad \Delta F^* \ \alpha (1/\Delta T)^2 \tag{1}$$

where ΔF^* is the critical free energy for nucleation.[6] It can be appreciated from Equation 1 that while ΔF^* is reduced by increasing ΔT, and hence N becomes greater, the lattice diffusion constant which controls particle growth is decreasing. If a fast cooling rate is maintained to below 1000°C, very small precipitates will result. Therefore, while initial undercooling should be rapid, this must be followed by a process which will allow coarsening of the precipitates so that at room temperature they are in a metastable state.

Three types of cooling sequences are most commonly used:

1. Rapid cool (quench), >500°C/hr
2. Controlled slow cool or sliding cool
3. Isothermal hold

The first two cooling sequences and their effects on microstructure will be discussed. The isothermal hold is a special heat-treatment schedule used for Mg-PSZ, the significance of this cycle will be described in Section IV.

1. Rapid Cool

Cooling faster than about 500°C/hr would be considered rapid. This type of cooling is impractical for large objects as it invariably leads to thermal shock fracture. At this cooling rate, the resulting precipitate size will be in the range of 30 to 60 nm (the precipitate morphology will depend upon the stabilizer type used).[7] These precipitates will have their M_s temperature well below room temperature and will require further heat treatment (aging) to grow the precipitates and bring them to a condition of metastability at room temperature. The temperature at which the aging treatment is carried out will be governed by a number of factors:

1. Practical or economic growth rates, i.e., do not want unrealistically long aging times, nor excessively high aging temperatures
2. Avoid eutectoid decomposition, as in ZrO_2-MgO system, where aging temperatures for rapidly cooled samples should not be <1400°C
3. Select aging times where equilibrium precipitate content is attainable, e.g., Figure 1a at AT, where the lever rule may be applied to determine the content as demonstrated earlier

Coarsening of the rapidly cooled specimens may be depicted schematically, as shown in Figure 3, in terms of the precipitate size distribution. In the rapidly cooled material all the particles will be below the critical size for stress-induced transformation (d_{STRESS}). The material is said to be under aged (UA). Aging, e.g., at 1300°C for Ca-PSZ and 1400°C for Mg-PSZ, causes the particles to grow and allows the amount of t-ZrO_2 to be optimized, or peak aged (PA). As discussed in Chapter 3, the volume fraction of transformable t-ZrO_2 is a critical parameter in determining the fracture toughness of the material. If the aging treatment is too long, e.g., >4 hr at 1400°C for Mg-PSZ, the precipitates will grow beyond the critical size, their M_s will be above room temperature, and they will spontaneously transform to monoclinic on cooling. This is termed the over aged (OA) condition.

It should be pointed out that it is only "static" or isothermal type heat treatments which allow the expected tetragonal precipitate content (according to the equilibrium phase diagram) to be realized. Continuous cooling will not allow the equilibrium content to be achieved.

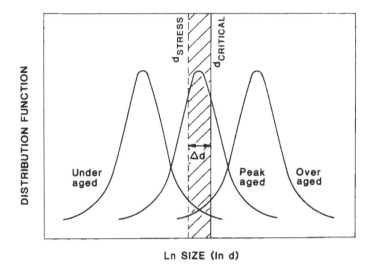

FIGURE 3.   Schematic illustration of the particle size distribution as a function
of the particle size for variously aged conditions. All particles to the right of $d_{critical}$
are monoclinic, whereas those on the left of d_{stress} will not transform on the
application of a stress. When the area with Δd is a maximum the material is said
to be peak aged (PA).

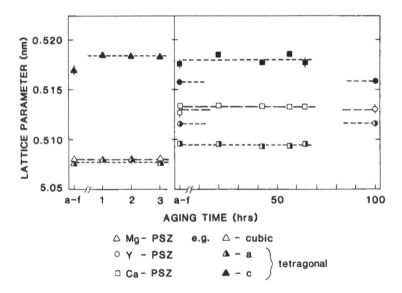

FIGURE 4.   Cubic and tetragonal lattice parameters of various PSZ alloys as a function
of aging time. Aging temperatures were 1400°C for Mg-PSZ and 1300°C for Y- and Ca-
PSZ. The constant nature of the lattice parameters, except for as-fired (a-f) value of Mg-
PSZ, indicates that the precipitated phase is at the equilibrium solute composition and
that precipitate growth does not influence the phase compositions at the aging temperatures
used.

As mentioned, the aging process should only coarsen (grow) the tetragonal precipitates
to critical metastability, and not cause additional precipitation other than that due to solute
content adjustment. Accurate lattice parameter measurements of the cubic and tetragonal
phases in "rapidly" cooled and aged Mg-, Ca-, and Y-PSZ materials indicate, with reference
to Figure 4, that:

1. The lattice parameters remain constant with aging time (the as-fired Mg-PSZ material does show a lower c_t lattice parameter, indicating perhaps that the material has not quite attained full "tetragonality"), i.e., a "high" Mg^{2+} solute content may still be present in the Zr lattice, such that the equilibrium tetragonal c/a ratio has not been achieved.
2. When there is low solid solubility of the solute at low temperature, as with MgO and CaO, the tetragonal lattice parameters for the differently stabilized materials are the same, cf. the c_t-axis for Mg- and Ca-PSZ. This feature indicates that very little solute is dissolved in the zirconia precipitates, unlike the precipitates in Y-PSZ materials.
3. The lattice parameter of the cubic stabilized zirconia (CSZ) matrix is directly related to the stabilizer used, but not necessarily related to the ionic radii ratio of the solute and solvent (see Chapter 2, Table 3). The CSZ lattice parameter also remains constant for the period of the aging times studied.
4. The lattice parameter difference between the cubic-stabilized phase, a_c, and the two tetragonal axes, a_t and c_t, is about equal for the Ca- and Y-PSZ materials, while for Mg-PSZ, the difference is about a factor of 20 times larger for the c-axis with respect to the cubic axis than it is for the tetragonal a-axis. It is considered that these axial differences in t-ZrO_2 lattice parameter in part account for the precipitate morphologies of the various precipitated tetragonal phases[7] (see Sections III, IV, and V).

2. Controlled Cool

An economically more appealing method of cooling the sintering/solution-treated bodies is by a controlled "slow" cool. These treatments were initially preferred, not as a matter of choice, but more of necessity, as cooling simply involved switching off the furnace. This process lead to early PSZ materials with very variable properties, generally OA because of the large thermal mass of the ceramic kilns.

Subsequent experiments have shown that PSZ alloys may be cooled at rates which produce materials at or near PA condition. Such cooling treatment can vary from continuous linear (sliding) to rate arrest and acceleration. The main purpose of the cool will be to have the body reside for some period of time in a region where the nucleation process has been completed and growth of the precipitated phase can occur. Continuous cooling generally produces materials with a range of precipitate sizes varying from grain to grain, and probably not having attained the equilibrium volume content. In addition, the thermal gradient existing through a large body may lead to a range of tetragonal precipitate sizes.

Therefore, an isothermal arrest during the cooling cycle is a better mode of cooling. This cooling technique has been studied intensively for the Mg-PSZ system and will be described in Section IV.

B. Fabrication of TZP

Starting powders for TZP ceramics generally consist of coprecipitated powders, i.e., zirconia plus alloy addition, in the size range of 10 to 200 nm and a composition of 1.5 to 3.5 mol % Y_2O_3.* Coprecipitated powders are favored, since these will ensure intimate mixing of the solute and solvent oxides, such that elemental homogenization takes place rapidly at moderate temperatures. Firing is carried out at 1300 to 1450°C in the tetragonal single phase field (see TZP region of Figure 1b). The final grain size is usually in the range of 0.5 to 2.0 μm and will depend upon firing time and temperature and solute content. The critical tetragonal grain size, within the compact, will depend upon solute content and compact density. It can also be appreciated why the lower firing temperatures of TZP ceramics are so beneficial, in that they naturally result in smaller grain size.

---

* Compositions in excess of 3.5 mol % Y_2O_3 would be fired in the c-phase field or near the two (t + c) phase-field boundary and would be considered as for PSZ materials as indicated in Section II.A.

Until recently, low firing temperatures meant that the theoretical density was difficult to obtain within a reasonable firing time. Therefore, a favored practice is to submit the presintered material with closed porosity to a short-time excursion up to ~1600°C (e.g., for ~15 min) in a hot isostatic press (HIP). This treatment promotes near-theoretical density without excessive grain growth. Higher quality powders that enable in excess of 98% of theoretical density to be achieved at 1400°C with a grain size <0.5 μm are now available. These materials are still often given an HIP excursion to increase strength and reliability.

III. Ca-PSZ

Commercial compositions of Ca-PSZ generally occur in the range 3 to 4.5 wt % CaO. Sintering and solution treatment are carried out as described in the hypothetical diagram of Figure 1a, with the firing temperature determined by the composition. Due to the low eutectoid temperature (see Chapter 2 Figure 18a), eutectoid decomposition is not a consideration in selection of a suitable aging temperature. Aging for commercial materials is carried out at 1300°C.[8] Figure 5 (a to c) shows the tetragonal precipitate coarsening sequence of an 8.4 mol% (3.8 wt %) Ca-PSZ alloy.

Hannink et al.[9] have plotted the precipitate growth rate, as a function of aging time at 1300°C, as shown in Figure 6. Their study showed that precipitate coarsening obeys a $t^{1/3}$ relationship as proposed by the Greenwood-Lifshitz-Slyzov-Wagner theory.[6] Transmission electron microscope observation and X-ray measurements showed that the precipitates remain predominantly tetragonal up to ~100 nm diameter, after which the precipitate M_s is above room temperature and the $t^{1/3}$ relationship breaks down, as indicated by the inflection in the plot of Figure 6. This precipitate value coincided with an aging time at which optimum mechanical properties were achieved, i.e., the PA material (see discussion Figure 8). Dilatometry was used to determine the M_s-M_f and A_s-A_f temperatures of the just-OA alloy studied by Hannink et al.[9] Test results are shown in Figure 7 from which it can be seen that the transformation temperature is a direct function of precipitate size and that for 100 nm, it is just below room temperature, the optimum condition for stress-assisted transformation. Marder et al. have also studied the coarsening kinetics of the tetragonal precipitates in Ca-PSZ.[10] Their work covered temperatures in the range of 1000 to 1400°C and their coarsening results confirmed those of Hannink et al.[9]

More recently Dickerson et al.[11] have measured the growth of t-ZrO_2, produced using commercial manufacturing thermal treatment. These workers found, that for t-ZrO_2 precipitates initially larger than 30 nm, the growth rate obeyed a $t^{1/2}$ relationship (a quadratic). This type of growth mechanism implies an interfacially controlled rate rather than by lattice diffusion as implied by a $t^{1/3}$ relationship.[6] The faster rate may come about when the equilibrium precipitate content has been achieved and the particles have impinged. The growth rate may then accelerate due to boundary diffusion at the contacting precipitates. A similar relationship ($t^{1/2}$) could be fitted to the precipitate size date >30 nm and for aging times >20 hr shown in Figure 6.

In quenched material (>1000°C/sec), the very small tetragonal precipitates are disk shape, as would be expected for precipitates with a small (<0.01) misfit between the particle and matrix.[12] After aging for ~10 hr at 1300°C, the precipitates adopt an irregular cuboid shape. The shape of the precipitates (in all PSZ systems) has been simplistically accounted for in terms of lattice parameter misfit.[7] Khatchaturyan[13] has explained the growth alignment of the tetragonal precipitates in terms of elastic strain effects, such that preferential alignment will occur in the "soft" elastic directions. Khatchaturyan's theory gives an excellent account for the tetragonal precipitate variant distribution in the Ca-PSZ. Lanteri et al.[14] have also used the theory to account for the morphology and "variant-banding" of the tetragonal precipitates in Ca-PSZ.

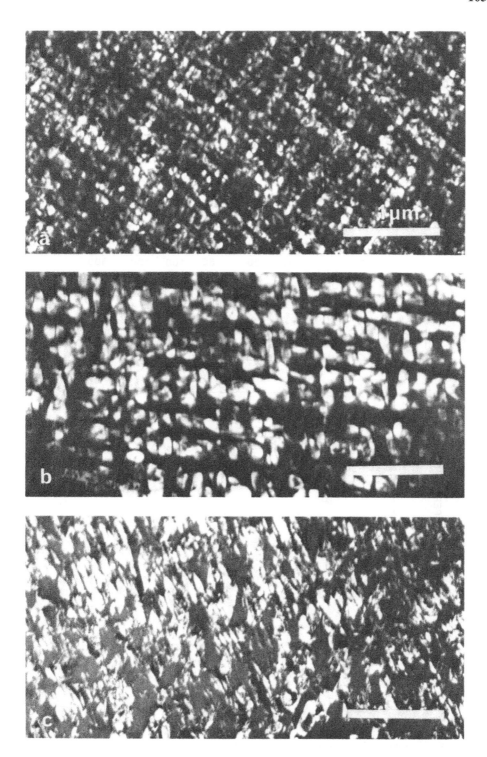

FIGURE 5. Dark field transmission electron microscope (TEM) images showing microstructural development of the precipitate phase in Ca-PSZ (a) as-fired, (b) peak-aged, and (c) over-aged. Bar length = 1 μm.

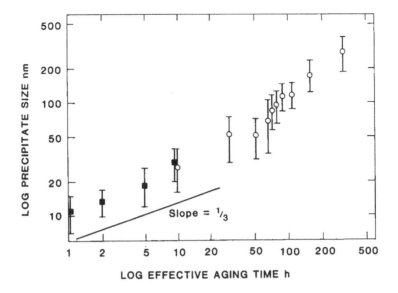

FIGURE 6.   Log plot of precipitate growth data; 10 hr was added to the aging times of the slow cooled samples to allow for growth during cooling. (After Hannink, R. H. J., Johnston, K. A., Pascoe, R. T., and Garvie, R. C., in *Advances in Ceramics,* Vol. 3, American Ceramics Society, Columbus, Ohio, 1981, 116.)

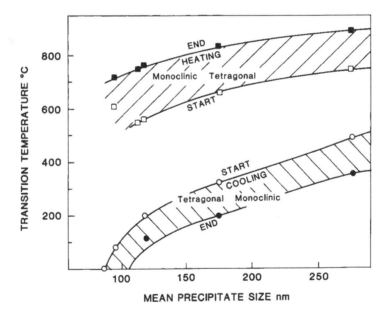

FIGURE 7.   Temperature for the start and finish of the reversible monoclinic-tetragonal transition as determined by dilatometer measurements. (After Hannink, R. H. J., Johnston, K. A., Pascoe, R. T., and Garvie, R. C., in *Advances in Ceramics,* Vol. 3, American Ceramics Society, Columbus, Ohio, 1981, 116.)

The tetragonal precipitate size and metastability is most strongly reflected in the mechanical properties. Figure 8 shows the modulus of rupture (flexural strength) as a function of aging time at 1300°C for a series of Ca-PSZ alloys. It can be seen from this figure that the peak strength coincides with a particular aging time, which is that time required to achieve the optimal amount and size of metastable tetragonal precipitates for the various alloy compo-

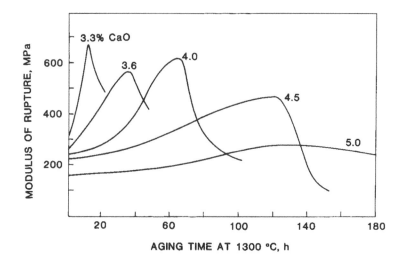

FIGURE 8. Flexural strength (MOR) attainable for various Ca-PSZ alloy compositions (weight % CaO) as a function of aging time at 1300°C. (After Garvie, R. C., Hannink, R. H. J., and Pascoe, R. T., U.S. Patent 4,067,745, 1978.)

sitions as a function of the initial cooling rate. Results also indicate the rapid drop off in strength as the samples are OA. The critical stress intensity (K_{IC}) and fracture surface energy can also be directly related to the precipitate size (metastability), as shown in Figure 9 (a and b).[15,16] In Figure 9a, the modulus of rupture (MOR) and fracture energy of a 3.8 wt % Ca-PSZ alloy are shown as a function of aging time. The dramatic increase in fracture energy, which can be directly related to K_{IC} through the relationship $K_{IC} = (2E\gamma_f)^{1/2}$ (where γ_f is the fracture surface energy and E the Youngs modulus) is one of the most significant benefits of transformation toughening. Similarly, K_{IC} can be directly related to the tetragonal precipitate size, as shown in Figure 9b.

It is also possible to determine the depth of transformation zones beneath a fracture surface. This information is required to determine the contribution of the transformation to the toughening increment (see Chapter 3). The technique used is one of X-ray absorption[17] or a two-wavelength technique.[18]* The results of such a determination, Figure 10, indicate that the volume fraction (V) and transformation zone depth (h) may be directly related to the fracture toughness contribution, in good agreement with the form of Chapter 3, Equation 11.

The influence of temperature on the K_{IC} of Ca-PSZ for two different aging times at 1300°C (different precipitate sizes) is shown in Figure 11. From this figure, the dramatic drop off in toughness and transformable monoclinic zirconia on the fracture surface is immediately apparent. The main reason for the loss in properties is that as the temperature is increased, the tetragonal phase becomes more stable and is, thus, less susceptible to stress-induced transformation (see Chapter 3, Section III.D).

IV. Mg-PSZ

Microstructural-mechanical property relationships are more complicated in this system, due to the diversity of phases and microstructural morphologies which may be generated as a function of thermal treatment. Compositions for commercial Mg-PSZ materials occur in the range of 8 to 10 mol % MgO. Firing takes place at temperatures just inside the cubic

---

* Transformation zone about a crack tip or partially cracked plate may be measured using Raman microprobe.[35]

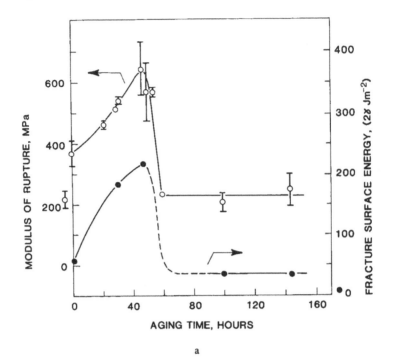

a

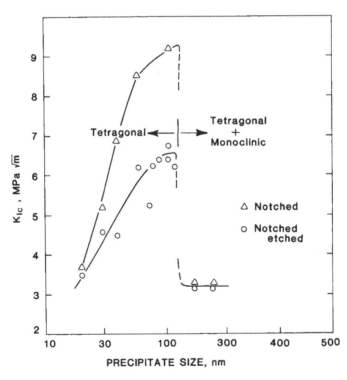

b

FIGURE 9.   (a) Flexural strength (MOR) and fracture surface energy plotted as a function of aging time at 1300°C for a 3.7 wt % Ca-PSZ alloy. (After Garvie, R. C., Hannink, R. H. J., and Urbani, C., *Ceramurgica Int.*, 6, 19, 1980.) (b) The influence of precipitate size on the critical stress intensity factor of Ca-PSZ. (After Swain, M. V., Hannink, R. H. J., and Garvie, R. C., in *Fracture Toughness of Ceramics*, Vol. 6, Bradt, R. C., Hasselman, D. P., and Lange, F. F., Eds., Plenum Press, New York, 1983, 339.)

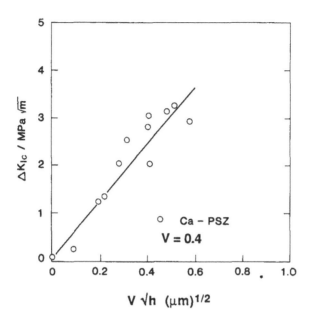

FIGURE 10. Increase in fracture toughness (ΔK_{IC}) for Ca-PSZ with transformation zone size h and volume fraction of tetragonal ZrO_2 (V) as anticipated from Chapter 3, Equation 11. (After Swain, M. V., Hannink, R. H. J., and Garvie, R. C., in *Fracture Toughness of Ceramics*, Vol. 6, Bradt, R. C., Hasselman, D. P., and Lange, F. F., Eds., Plenum Press, New York, 1983, 339.)

phase field (see Chapter 2, Figure 19) after which the materials are generally cooled, according to the two methods described in Sections II.A.1 and II.A.2, viz. rapid or controlled cool.

Cooling rates on the order of 500°C/hr are considered rapid for a 9.7 mol % Mg-PSZ alloy. The resulting precipitate size will be in the range of 30 to 60 nm and, as described in Chapter 2, Section IV.F, occur as small ellipsoidal- or lens-shaped particles with $(100)_c$ habit planes with the c_t-axis always parallel to the axis of rotation of the lens. These precipitates require a further aging treatment to bring their M_s to near room temperature. A number of aging treatments are possible; however, for materials with precipitate sizes below about 120 nm diameter, only aging temperatures above the eutectoid temperature (1400°C, see Chapter 2, Figure 19) should be considered. This is because the eutectoid decomposition reaction will dominate before the coarsening step brings the precipitates to a PA condition. We shall see later how "subeutectoid aging" of appropriately sized precipitates can produce very beneficial mechanical properties. Figure 12 shows the coarsening sequence of small precipitates following an aging treatment at 1420°C. Optimal or PA tetragonal precipitates have diameters of ∼180 nm and a thickness of ∼40 nm.

As described in Section II.A.1, controlled cooling is an economically better way of cooling PSZ materials. By employing this method, materials with maximum strength may be produced from a single firing. Optimum precipitate size from such a cool should be a little smaller than the PA condition, so that the material may be subeutectoid aged to achieve maximum fracture toughness.

Hughan and Hannink[19] have made a detailed study of the precipitation process in a 9.7 mol % Mg-PSZ material using isothermal hold steps in a 500°C/hr cooling curve which started at 1700°C. The isothermal holds were carried out for various periods of time. From such a cooling cycle, three different precipitate formation sequences were identified. These precipitates could be described as primary, large random, and secondary precipitate growth.

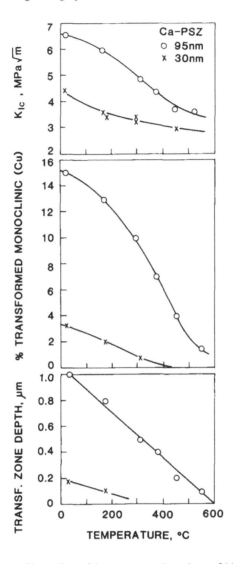

FIGURE 11.   Observations of the temperature dependence of (a) K_{IC}, (b) % of transformed monoclinic on the fracture surface (Cu radiation), and (c) the depth of the transformed zone for Ca-PSZ. Two sets of data are presented (o) and (x) for a material with 95 nm and 30 nm precipitate particles, respectively. (After Swain, M. V., Hannink, R. H. J., and Garvie, R. C., in *Fracture Toughness of Ceramics*, Vol. 6, Bradt, R. C., Hasselman, D. P., and Lange, F. F., Eds., Plenum Press, New York, 1983, 339.)

Primary precipitates were similar to those ascribed to the rapid cooling cycle (see Figure 12a). These precipitates nucleate homogeneously below the c/(c + t) phase boundary resulting from the supersaturation of the cubic solution, as mentioned in Section II.

Large precipitates are observed in materials continuously cooled to room temperature at 500°C/hr, but are more numerous in material isothermally held at temperatures above 1400°C. These precipitates nucleate at matrix inhomogeneities, such as pores or inclusions, and grow into clusters by what is assumed to be autocatalytic nucleation; they are all monoclinic at room temperature.

Secondary precipitate growth (SPG) occurs rapidly in a temperature zone of about 1300

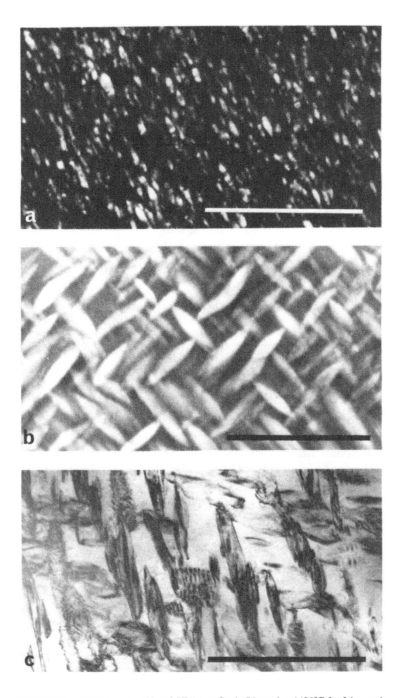

FIGURE 12. TEM images of Mg-PSZ (a) as-fired, (b) aged at 1420°C for 2 hr, peak-aged (PA), and (c) for 4 hr, over-aged (OA). (a) Dark field image and (b) and (c) bright field images, bar lengths equals 0.5 μm.

to 1400°C, just below the eutectoid temperature and 350°C below the c/(c + t) phase boundary. SPC appears to originate within a grain, but near grain boundaries, from favorably oriented primary precipitates. SPG does not necessarily proceed to completion by elimination of all the primary precipitates in the grains. When SPG has not gone to completion, polished and etched sections of the material exhibit a spot or mottled type contrast, as shown in

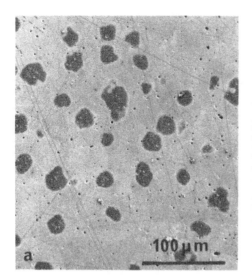

FIGURE 13.   Scanning electron microscope (SEM) images of (a) "spot" type contrast observed in Mg-PSZ held for 40 min at 1350°C. (b) Shows the three tetragonal variants in the secondary precipitate growth region while the small precipitate result from the primary precipitation process. The "black" discs at the bottom left hand corner are where in-plane precipitates have been removed as a result of the etching treatment. (After Hughan, R. R. and Hannink, R. H. J., *J. Am. Ceram. Soc.*, 65, 556, 1986.)

Figure 13a. Higher magnification scanning electron micrograph images of the spot region show the presence of two precipitate sizes, those of the primary within the spot and the secondary in the peripheral regions (Figure 13b). Hughan and Hannink[19] were able to show, by means of X-ray diffraction (XRD) and strength measurements, that when optimally grown, in terms of precipitate size (temperature dependent) and volume fraction (hold-time dependent), SPG provides the bulk of transformable particles in the enhanced mechanical properties of Mg-PSZ.

Secondary precipitate growth results in precipitates which are fairly uniform in size, related to the growth temperature, and once grown to that size, are remarkably stable against further growth.[20] An example of this stability is shown, in Figure 14, for an experimental Mg-PSZ material. Figure 14a shows secondary precipitates and some large random precipitate growth within "spots", after a 140 min isothermal hold at 1375°C. The same material isothermally held for 1290 min at 1375°C shows the "spot" area to be almost entirely consumed by large, random precipitates, while the precipitates in the outer zone have not grown (Figure 14b).

The phenomenon of stabilization of precipitates against coarsening has been considered for metals by Brown et al.[21] and Johnson.[22] Brown et al. showed that an array of tetragonal precipitates can become stable against further coarsening due to the interaction of the elastic strain fields between the precipitates and the balancing interfacial energy. Brown et al. model[21] thus proposed precipitate stability if:

$$\frac{2r\mu f\epsilon^2}{\gamma} > 5 \tag{2}$$

where r is the precipitate radius; μ, the shear modulus; f, the volume fraction of precipitates; ϵ, the constrained elastic misfit; and γ, the interfacial energy. While the equation can account for particle stability, experiments with Mg-PSZ have shown[19] that the precipitate size is a

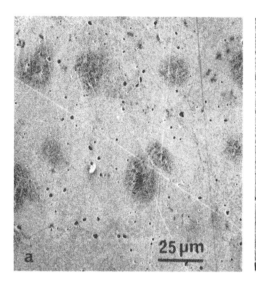 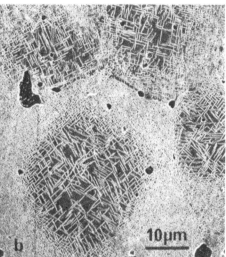

FIGURE 14.   SEM images of precipitates in Mg-PSZ after an isothermal hold at 1375°C for (a) 140 min and (b) 1290 min. Note how secondary precipitates have not grown significantly while the precipitates within the spot have now grown larger than the outer areas.

sensitive function of isothermal hold temperature. Thus, it is difficult to use the relationship to predict the equilibrium size at a particular temperature since the parameters are not well known as a function of temperature.

A feature not well understood about SPG is the cause of the slowdown in growth as the precipitates grow into the grain center, as demonstrated in Figure 14. Possible reasons may include a lack of strain or chemical energy as the growth proceeds away from the grain boundary regions. The driving force is sufficient, however, for complete primary precipitate conversion if a controlled sliding cool is used in the region of 1400 to 1200°C.

A. Subeutectoid Aging

It has been demonstrated that a subeutectoid aging treatment can produce one of the toughest sintered ceramic materials yet produced.[23-25] For suitably prefired Mg-PSZ, the best aging temperature is 1100°C. Suitable prefired materials are those which contain near optimally sized precipitates and are most readily produced by either a controlled or isothermal-hold cooling sequence.

The microstructural influence, hence mechanical properties, of the aging treatment will also depend on the precipitate size of the prefired materials, and these may be summarized into four main features.[26] First, the anticipated decomposition reaction, according to the phase diagram (Chapter 2, Figure 19), cubic-ZrO_2 to monoclinic-ZrO_2 plus MgO occurs at the grain boundaries and around pores. While the reaction is rapid in fully stabilized zirconia (14 mol % MgO),[27] it is not significant for the aging times used in PSZ. The reaction may also be considerably slowed down with the addition of suitable sintering aids[3] or minor additions of stabilizer. There are three other processes which may occur within the grains:

1.   The formation of an ordered anion vacancy phase, $Mg_2Zr_5O_{12}$ (δ-phase) at the precipitate matrix interface.[28]
2.   The precipitation of very small tetragonal precipitates within the cubic matrix of the precipitate-laden grains
3.   The transformation on cooling (M_s above room temperature) of some of the original tetragonal precipitates

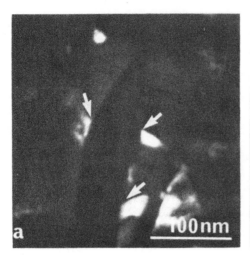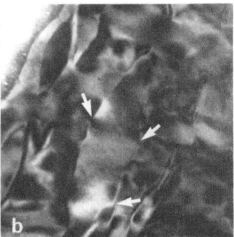

FIGURE 15.   TEM images of subeutectoid aged Mg-PSZ. (a) Dark field image, δ-phase precipitates at the t-ZrO₂/cubic matrix interface. (b) Bright field image showing strain within t-ZrO₂ precipitate resulting from the presence of the precipitates shown in (a).

For aging times used, the original precipitates show no significant increase in size. The main cause of the transformation to the room temperature form is the additional strain induced into the particle interface as a result of the δ-phase precipitation.[26]

Transmission electron microscope observations can reveal the development of strain within the precipitates. Figure 15 shows the strain centers developed within a precipitate when two δ-phase variants impinge at the tetragonal precipitate interface. These highly strained sites are thus favorable locations to initiate the transformation on the application of stress.

Depending upon the aging time at 1100°C, varying mechanical properties may result (see Section IV.C). The main difference in the microstructural aspects between aging times, and hence the different properties, is the amount of original tetragonal converted to monoclinic on cooling. The formation of this monoclinic consists of very small particles, 150 to 200 nm in diameter, twinned on a fine scale. The remaining tetragonal particles are exceedingly metastable such that their nucleation free energy, ΔF^*, is virtually zero. These particles will transform as a result of thermal stresses induced by an electron beam.[29] Similarly, their extreme sensitivity to applied stresses results in very large process zones and a rising crack resistance, R-curve, behavior for a propagating crack (see Chapter 3).

At higher subeutectoid aging temperatures, more than 1100°C, the decomposition kinetics occur at a more rapid rate.[30] The consequences are the formation of a thick, decomposed region at the grain boundaries within which microcracking and crack extension occur. The reason for the microcracking is that there is a substantial difference in the thermal expansion coefficient of the matrix and the grain boundary decomposed phase. This leads to a state of tension within the grain boundary phase on cooling, and when stressed, this phase will fracture.

B. Thermal Expansion Effects

The thermal expansion of PSZ materials is a very sensitive function of monoclinic phase content. This is due to the considerable thermal expansion difference between the tetragonal ($\sim 9 \times 10^{-6}$ per °C) and the monoclinic ($\sim 6 \times 10^{-6}$ per °C) phases. There is also hysteresis in the thermal expansion of the materials when the transformation occurs. In the upper portion of Figure 16 the thermal expansion data are shown for a variety of aging times at

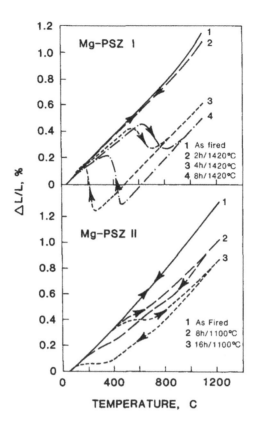

FIGURE 16. Normalized thermal expansion curves for Mg-PSZ follow-ing two types of heat treatment. Arrows indicate heating and cooling curves. (After Hannink, R. H. J. and Swain, M. V., in *Tailoring of Multiphase and Composite Ceramics Conference*, Messing, G. L., Pan-tano, C. G., and Newnham, R. E., Eds., Plenum Press, New York, 1986, 259.)

1400°C of a 9.7 mol% Mg-PSZ alloy. It can be seen from the data that virtually no inflection (i.e., no transformation) occurs until the material approaches the OA condition. The inflection is quite sharp, and its actual temperature is a sensitive function of the aging time at 1400°C.

Thermal expansion data for the 1100°C aged materials, starting with relatively small tetragonal-ZrO_2 precipitates, are shown in the lower portion of Figure 16. The expansion behavior of these materials is considerably different from those of the 1400°C aging sequence. The form of the curve depends upon the initial precipitate size. The temperature at which the transformation commences is generally lower than the 1400°C aged material, as a result of the smaller monoclinic precipitate size. The back transformation temperature is also a function of the size of the precipitates participating in the transformation, and is low even after 16 hr at 1100°C (see lower portion of Figure 16). After prolonged heating at either temperature, 1100 or 1400°C, a second major inflection is observed at about 1200°C. This inflection occurs as a result of the coarse monoclinic grain boundary phase transformation.

C. Thermomechanical Properties

1. Strength

The strength of Mg-PSZ, like that of other PSZ systems, varies with stabilizer composition, e.g., see Figure 8. The influence of composition on the flexural strength of proeutectoid-aged (1400°C)[31] Mg-PSZ is shown in Figure 17 (toughness varies in a similar manner with

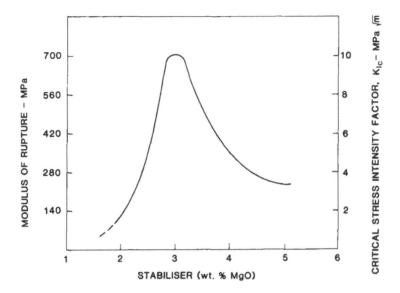

FIGURE 17.   Flexural strength and fracture toughness behavior of various Mg-PSZ alloys. (After Dworak, U., Olapinski, H., and Thamerus, G., *Sci. Cer.*, 9, 543, 1977.)

composition). The peak fracture properties occur when the percentage of metastable phase is ~50%. Because of the peak properties attainable, and the greater control over the precipitate coarsening kinetics, stabilizer contents in the range of 3.3 to 3.4 wt % MgO are now preferred for commercial alloys.

The effect of aging temperature on four-point bend strength of ground bars in a 3.3 wt % Mg-PSZ alloy, given aging treatments above and below the eutectoid temperature for various times, is shown in Figure 18.[32] The aging of these samples commenced with the "as-fired and cooled" precipitate sizes. It can be seen from this figure that the 8 hr age at 1100°C yields the best value.

The influence of pre-aged precipitate size on subeutectoid aging behavior is demonstrated in the series of curves in Figure 19 (a and b). The family of curves in Figure 19a shows the maximum strength occurring after longer or shorter aging, depending upon the initial precipitate size, the important feature being that only one size, ~180 nm in the long dimension, yields the optimal strength. Figure 19b shows the monoclinic-ZrO_2 content on the ground surfaces of the test bars. As discussed in Chapter 2, Section III, when the precipitates are too small, they are not readily transformed, e.g. by grinding or by stress at a crack tip. The precipitates become more metastable when aged, leading to improvements in strength and toughness. If, however, the 1100°C aging treatment is too long, decomposition and coarsening of the monoclinic grain boundary phase will lead to a degradation in the properties.[25] If the tetragonal precipitates are initially too metastable, then short aging times will lead to a destabilization and a corresponding reduction in properties.

The technique of controlled cooling interrupted by isothermal holds may yield very good strength properties from the "as-fired" condition. Figure 20 shows the flexural strength attainable as a function of hold temperature and time. The peak value of ~650 MPa attained after 120 min hold at 1340°C indicates the optimum secondary precipitate growth conditions. The flexural strength values obtained by this technique are comparable to the best "fired-and-aged" strengths attainable.[19]

As with all PSZ materials, the strength of Mg-PSZ decreases with increasing test temperature because of the increasing stability of the tetragonal-ZrO_2 precipitates. The stability corresponds to an increase in the ΔF^* and a reduction in the chemical driving force, namely

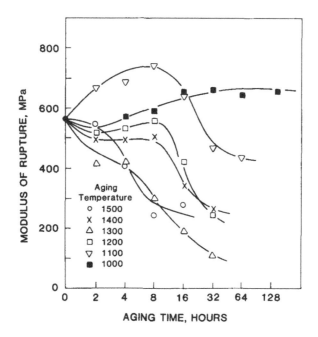

FIGURE 18. Effect of heat treatment time and temperature on the flexural strength (MOR) of Mg-PSZ containing 9.4 mol % MgO. (After Swain, M. V., Hughan, R. R., Hannink, R. H. J., and Garvie, R. C., *Uehida Rokakuho*, 2, 117, 1984.)

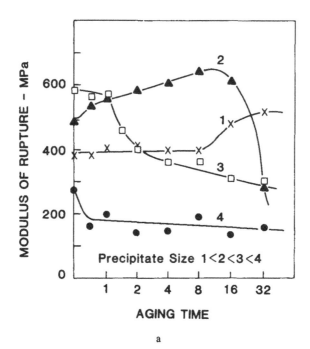

a

FIGURE 19. Effect of initial precipitate size on (a) the flexural strength (MOR) and (b) monoclinic content of an Mg-PSZ alloy containing 9.4 mol % MgO when aged at 1100°C for various times. (After Swain, M. V., Hughan, R. R., Hannink, R. H. J., and Garvie, R. C., *Uehida Rokakuho*, 2, 117, 1984.)

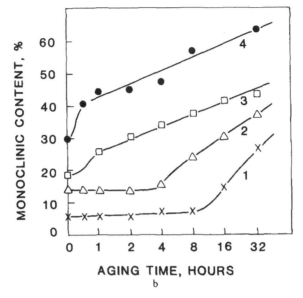

FIGURE 19 (continued)

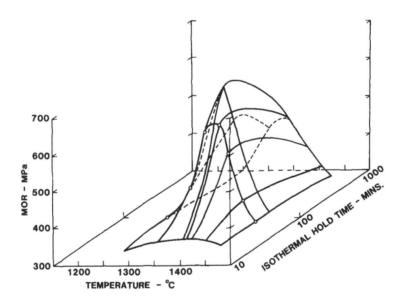

FIGURE 20.   A diagrammatic illustration of the flexural strength (MOR) as a function of isothermal hold temperature and time. The 80-min-hold-time data is shown. (After Hughan, R. R. and Hannink, R. H. J., *J. Am. Ceram. Soc.*, 65, 556, 1986.)

ΔF_{chem} in Chapter 2, Figure 6b. Typically at temperatures of only 500 to 600°C, the strength is half the room temperature value, but remains constant thereafter to 1100°C.

2. Fracture Toughness

The fracture toughness of Mg-PSZ, like the strength, is a function of composition, precipitate volume and size, and heat treatment. Figure 21 (a to c) illustrates the change in fracture surface roughness as a function of precipitate development. In these materials, the fracture toughness has increased from (a) ~2.5 MPa m$^{1/2}$ to (c) ~8.5 MPa m$^{1/2}$. The reason

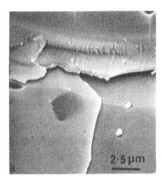

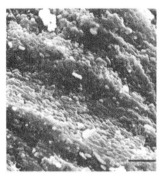

FIGURE 21. Scanning electron microscope (SEM) images of the fracture surfaces of an Mg-PSZ alloy as a function of increased K_{Ic}. (a) $K_{Ic} = \sim 2.5$ MPa.m$^{1/2}$, (b) $K_{Ic} = \sim 4$ MPa.m$^{1/2}$ and (c) $K_{IC} = \sim 8.5$ MPa.m$^{1/2}$.

for the rougher surface can be attributed to both the greater crack deflection caused by the transforming precipitates and to the transformation-toughening component.

The optimum toughness is more readily and reliably achieved by subeutectoid aging rather than proeutectoid aging treatments. Figure 22 (a and b) compares the K_{IC} and fracture surface energy following the two different aging treatments; the greatly enhanced fracture energy of the 1100°C aged material is one of its impressive benefits. This material, at the peak values, displays crack growth stability during notched fracture tests, indicative of R-curve behavior (Chapter 3), whereas the 1400°C aged material behaves in a more conventional brittle manner.[33]

Variations in the initial precipitate size and aging times at 1100°C enables K_{IC} to be varied in the range of 3 to 15 MPa$\sqrt{m}$. The lower K_{IC} values correspond to the case where the precipitates are too small to be transformed by the crack tip stress; at the other extreme, the critical stress to transform the tetragonal precipitates is comparable to the bend strength and the transformation zones may extend up to 50 μm from the crack tip.[33] Such data may be used to test the viability of equations such as Chapter 3, Equation 11 equating the toughness increment to the volume fraction of transforming precipitates, zone size, etc. The results of Figure 23 confirm the toughness increment dependence as the square root of the zone size.[34]

The fracture toughness, like strength, decreases with increasing temperature. This effect is shown in Figure 24 for the same Mg-PSZ material having received five different aging treatments. The 8 hr/1400°C material contains predominantly monoclinic-ZrO$_2$ precipitates, whereas the other materials were essentially tetragonal. As mentioned previously for Ca-PSZ, Figure 11, the transformation zone size (depth) decreases rapidly with temperature, thereby accounting for the decrease in toughness.[16] The reduction is most severe in the proeutectoid aged materials and least in the "OA subeutectoid" material. The explanation given, in terms of the thermodynamic stability for the strength, also pertains to the fracture toughness. A more systematic method of expressing the toughness of Mg-PSZ (and other PSZ materials) is in terms of the M_s temperature. Becher et al.[35] examined a range of Mg-PSZ materials to determine M_s by sub-zero cooling. They found that the toughness varied linearly with M_s temperature (see Figure 25). It was also found that for a specific material, the slope of the K_{IC}-temperature curve shifted, depending upon the M_s temperature. This is shown in Figure 26 for two Mg-PSZ materials with tetragonal-ZrO$_2$ precipitates of different size.

R-curve behavior is known for PSZ materials but is most marked in the subeutectoid materials. This behavior has been described in Chapter 3, Section III.E. Another improvement in the mechanical properties resulting from suitable aging treatments is that of thermal shock resistance.[24,36]

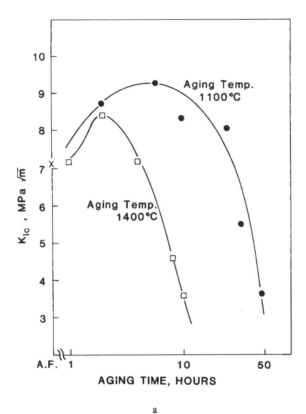

a

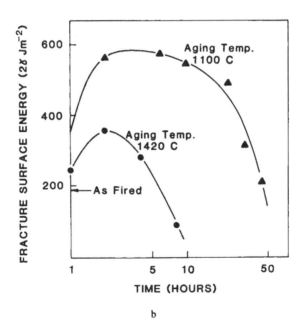

b

FIGURE 22.   Comparison of fracture behavior of Mg-PSZ aged at two different temperatures (a) K_{IC} measured by notch beam test and (b) fracture surface energy. (After Hannink, R. H. J. and Swain, M. V., in *Tailoring of Multiphase and Composite Ceramics Conference*, Messing, G. L., Pantano, C. G., and Newnham, R. E., Eds., Plenum Press, New York, 1986, 259.)

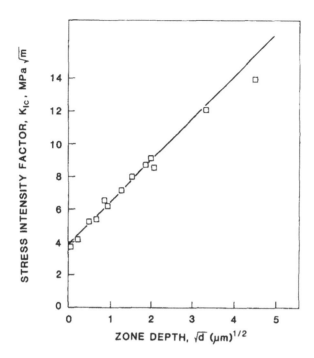

FIGURE 23. Relation between transformation zone depth and critical stress intensity factor for Mg-PSZ materials containing 9.4 mol % MgO. (After Swain, M. V., *Acta Metall.*, 33, 2083, 1985.)

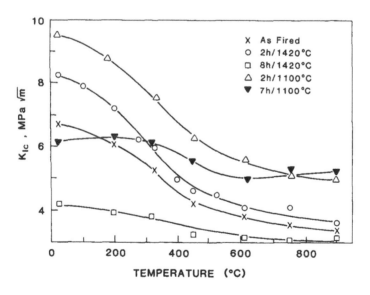

FIGURE 24. Temperature dependence of critical stress intensity factor, K_{IC}, in specimens of Mg-PSZ, given various prior heat treatments. (After Hannink, R. H. J. and Swain, M. V., in *Tailoring of Multiphase and Composite Ceramics Conference*, Messing, G. L., Pantano, C. G., and Newnham, R. E., Eds., Plenum Press, New York, 1986, 259.)

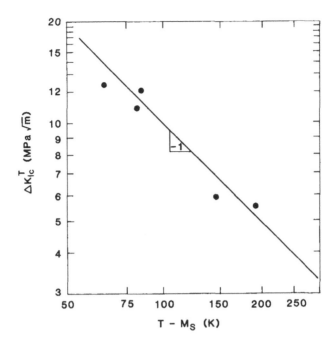

FIGURE 25.   The transformation-toughness contribution, ΔK_{IC}, increases as the (T-M_s) decreases. The transformation-toughening contribution is inversely proportional to (T-M_s) for the case where T = 22°C > M_s. Note K_{IC} has been normalized to constant volume fraction (V) in order to eliminate V variations in the test samples. (After Becher, P. F., Swain, M. V., and Ferber, M. K., *J. Mater. Sci.*, 22, 76, 1987.)

As seen in Chapter 3, apart from transformation toughening, there are other mechanisms which contribute to the toughness of PSZ alloys; these include crack deflection and crack bridging, due to the presence of the highly ellipsoidal precipitates (see Figure 12), and the possibility of microcracking at the matrix-precipitate interface. All these toughening mechanisms will contribute to the matrix toughness, to which the transformation-toughening increment is added. Examples of crack deflection and crack bridging about coarse precipitates in proeutectoid OA Mg-PSZ are shown in Figure 27. The extreme tortuosity, coupled with crack branching and precipitate bridging behind the crack tip, would also contribute to R-curve behavior in this material. Similarly, microcracking in the monoclinic ZrO_2 grain-boundary phase of subeutectoid-aged materials also contributes to the R-curve behavior.[33]

V. Mg/Ca-PSZ

Mg/Ca-PSZ has been used as wear-resistant material since 1975, although not in a transformation-toughening form.[37] Total stabilizer content is 3 wt % MgO-CaO. Assuming Vegard's Law* for the lattice parameter of the cubic matrix phase (a_0 = 0.50920 nm), the Mg to Ca composition ratio is 50:50 atomic percent of total stabilizer content. This composition would yield an equivalent stabilizer content of 9.175 mol %, which is about the ideal commercial composition for "insoluble" type stabilizer materials, with a eutectoid composition at ~14 mol %, in terms of producing the optimum volume fraction of tetragonal phase for transformation toughening purposes.

It is stated that for these materials, to produce wear-resistant materials, thermal treatment

---

* Vegard's Law assumes that the lattice parameter of a solid solution varies linearly with atomic concentration.

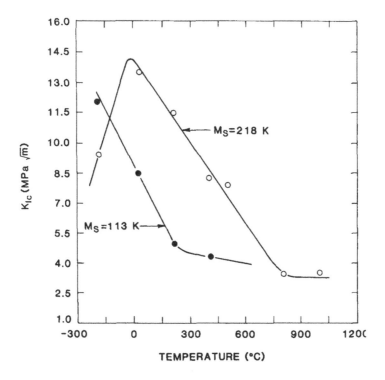

FIGURE 26. From thermodynamic considerations of transformation toughening, the critical fracture toughness will be a function of test temperature (T). With T > M_s, K_{IC} decreases with increasing T for Mg-PSZ with a given M_s because of increased tetragonal stability as T is raised. When T < M_s, a proportion of the t-ZrO$_2$ will spontaneously transform prior to testing, hence V decreases and K_{IC} is lowered. Raising the materials M_s temperature, shifts the K_{IC} peak to a higher test temperature. (After Becher, P. F., Swain, M. V., and Ferber, M. K., *J. Mater. Sci.*, 22, 76, 1987.)

is carried out so that the resultant phase ratios are in the range of 90 to 95% cubic, 5 to 10%, tetragonal, and <5% monoclinic; the aim of the heat treatments being to achieve maximum hardness.[38] The grain size, after firing, is about 60 μm. This size is comparable to those of the Ca- and Mg-PSZ sintered/solution treated materials.

Transmission electron microscope observations reveal that the tetragonal precipitate morphology is variable throughout the material, at times appearing like that of Ca-PSZ and at others like Mg-PSZ. This variability suggests an inhomogeneous distribution of the stabilizing cations. The as-fired precipitates are irregular-equiaxed shape and they appear similar to those of Ca-PSZ, cf. Figure 5. After aging at 1400°C, the precipitate morphology adopts a form somewhat in between that of its two parent stabilizers, a lozenge form. Insufficient data is available to determine the equilibrium aspect ratio.

While long-term, low-temperature aging data is not available, it is known that alloys with low eutectoid temperatures, when added to Mg-PSZ, will reduce the decomposition kinetics (see Section IV). Thus, high-temperature stability would be one of the benefits expected of Mg/Ca-PSZ alloys.

Mechanical property data is also limited for this system. For most of the data that has been published, thermal treatment of the material and the metastable condition of the tetragonal phase was not stated. Li and Dworak[39] have reported mechanical properties of a Mg/Ca-PSZ alloys; their results are shown in Table 2.

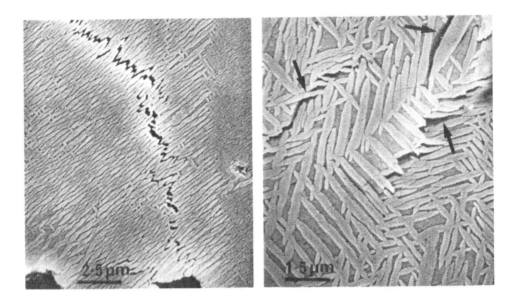

FIGURE 27.   The tortuous crack path in over-aged proeutectoid heat-treated Mg-PSZ showing precipitated matrix delamination and crack bridging (arrowed). (After Hannink, R. H. J. and Swain, M. V., *Tailoring of Multiphase and Composite Ceramics Conference,* Messing, G. L., Pantano, C. G., and Newnham, R. E., Eds., Plenum Press, New York, 1986, 259.)

Table 2
FRACTURE AND HARDNESS
PROPERTIES OF A Mg/Ca-PSZ,[a]
MEASURED AFTER VARIOUS
TREATMENTS

Property

K_{IC}, as-fired, RT measured	4.6	MN m$^{-3/2}$
K_{IC}, 1000°C	2.5	MN m$^{-3/2}$
K_{Ic}, after 1500°C anneal	3.5	MN m$^{-3/2}$
Bend strength	350	MPa
Vicker's hardness	15	GPa

[a]   Feldmuhle ZT35, Feldmuhle AG.

VI. Y-PSZ

Y_2O_3-ZrO_2 materials occupy a broad spectrum of compositions and properties in ZrO_2 alloy materials. Y-PSZ materials, as defined in the introduction of this chapter, are derived from a high-temperature solid/solution treatment followed by a controlled precipitation process. In our description of Y-PSZ we shall adhere to this definition of the materials.

A limited amount of single-crystal and polycrystalline data is available on Y-PSZ. Considerable difference has been reported in the microstructures and mechanical properties attainable[40-42] (see also Section VII). These differences are, in part, due to the sluggish nature with which these materials approach equilibrium, particularly when cooled from the melt (see Section VII).

As shown in Figure 1, Y-PSZ compositions occur in the range of 3 to 6 mol % (6 to 12.5 wt %) Y_2O_3. At these compositions, solution treatment in the cubic phase field is uneconomic. Therefore solution treatment is often dictated by the firing facility available

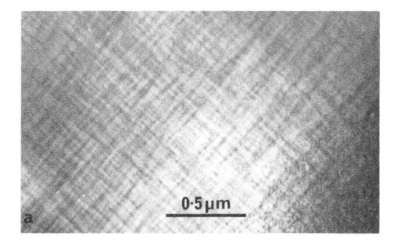

FIGURE 28. Y-PSZ (a) as-fired "tweed" microstructure, transmission electron microscope (TEM) bright field image, and (b) dark field image of material aged 100 hr at 1300°C.

and is generally in the range of 1750 to 2000°C. The fired grain size is large, usually 50 to 70 μm. Firing is followed by rapid cooling and an aging schedule; controlled cooling and isothermal holds during cooling have not been reported for these materials. Figure 28a shows the as-fired microstructure of rapidly cooled material and has the characteristic tweed form of contrast. The tweed contrast occurs due to a "strain induced coarsening in coherent mixtures of cubic and tetragonal phases."[13] The tweed or basketweave structure occurs in all rapidly cooled PSZ materials; the banding in Figure 5 is a consequence of the same phenomenon. Figure 28b shows the coarsened precipitate morphology after 100 hr at 1300°C where the precipitates have grown and agglomerated into rectangular plates or colonies composed of tetragonal twin-related variants.[43]

Lanteri et al.[43] have made a detailed study of the tetragonal phase in a single crystal 8 wt % Y_2O_3-ZrO_2 alloy.[9] These workers observed two tetragonal forms (t and t') in material grown from the melt. After annealing at 1600°C, the t'-phase decomposed into a low-solute t-phase and a high-solute cubic phase. It is only when the fabricated materials approach a stage of equilibrium, such that the low solute t-phase is present, that transformation toughening can be induced.

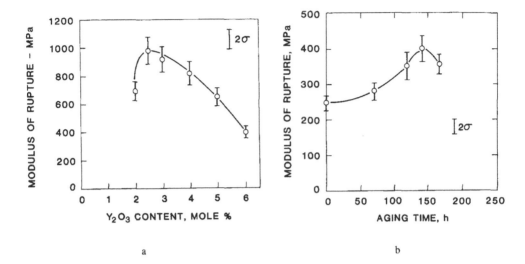

FIGURE 29.   (a) Effect of Y_2O_3 content on flexural strength of Y-PSZ materials. (b) Change in flexural strength during aging of 4.5 mol % Y-PSZ composition at 1400°C for t-phase precipitation. (After Ingel, R. P., Ph.D. thesis, 1982.)

The PA strength of Y-PSZ materials can approach those of other PSZ systems. As seen in Figure 29a, the strength decreases with increasing Y_2O_3 content (similarly the fracture toughness). Matsui et al.[44] have compared the mechanical properties achievable from Y_2O_3-ZrO_2 in the range of 2.5 to 6 mol ($\sim$4 to 10 wt %) Y_2O_3.[44] These workers found a strength composition peak at 2.5 mol % ($\sim$5 wt %) Y_2O_3.

Figure 29b shows the aging data acquired from a 4.5 mol ($\sim$9 wt %) Y-PSZ alloy aged at 1400°C; the strength attainable was 400 MPa after aging 140 hr.[44] This peak strength was attributed to the coarsening tetragonal precipitates, rather than transformation toughening, while the low value of the peak strength was linked to the large cubic grain size and low toughness.

The shape of the curve in Figure 29a and the long aging times of Figure 29b indicate the potential benefits of the lower Y_2O_3 content with the accompanying promise of higher strengths and lower grain sizes. Therefore, to produce material with better mechanical properties, Y-TZP rather than Y-PSZ is used, as discussed in Section VIII.

VII. PSZ SINGLE CRYSTALS

Single crystals of PSZ have been made by numerous single-crystal growing techniques, varying from flux growth of m-ZrO_2 crystals at low temperatures (<1150°C) to skull melting and floating zone method for t- and c-ZrO_2 crystals.[42] The major enterprise in this area has been the skull melting of c-ZrO_2 for use as diamond simulant for the jewelry industry. This technique has been successful in fabricating large single crystals of a number of refractory oxides.[45] The floating zone technique is suitable for fabricating thin, single-crystal rods of various materials. Examples of the two crystal-growing techniques are shown in Figures 30 and 31.

The number of studies referring to mechanical properties of PSZ single crystals is limited.[40,46-48] Ingel[40] studied the influence of Y_2O_3 content on the mechanical properties of Y-PSZ single crystals over the composition range 0 to 20 $^w/_o$ (wt %) Y_2O_3 produced by the skull melt technique. Limited studies of single crystals containing HfO_2 partially stabilized with Y_2O_3, two Ca-PSZ (4 and 8 $^w/_o$) compositions, and 2.8 $^w/_o$ Mg-PSZ. Michel et al.[47,48]

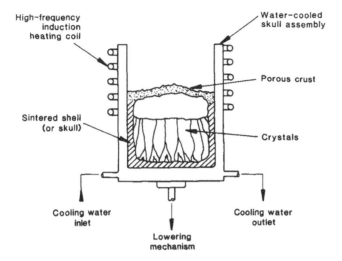

FIGURE 30. Schematic experimental arrangement for the skull-melting technique of single crystal growth.

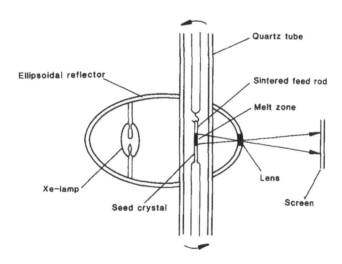

FIGURE 31. Floating zone method with a Xenon arc image furnace arrangement.

prepared smaller skull melted crystals of Y-PSZ containing 3 and 9 mol % Y_2O_3. These specimens were cooled at $> 400°C$/hr such that the 3 $^m/_o$ (mol %) material was single-phase t-ZrO_2 stabilized. Previous studies by Lefevre et al.[49] had established that single-phase t-ZrO_2 could be prepared by rapid-quenching ZrO_2 stabilized with M_2O_3 oxides (M = Sc, Y, or lanthoxides) in a narrow composition range (2 to 6 $^m/_o$) from $>2000°C$.

Mizutani, Kato, and co-workers[42,46] prepared a range of Y- and Sc-ZrO_2 single-crystal materials by the floating zone technique. Mechanical properties were only reported for the Y-PSZ materials over the composition range 1 to 8 mol % Y_2O_3. Electrical resistivity of both stabilized systems were measured over the temperature range 400 to 1200°C. Other studies of single-crystal properties include the creep deformation of Ca-CSZ[50] and indentation hardness anistotropy of Ca and Y-CSZ.[51]

a

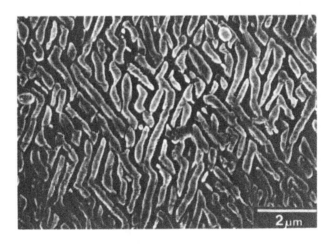

b

FIGURE 32.   SEM observations of the microstructure of chemically etched (H_2PO_4) single crystal Y-PSZ — single crystals prepared by the floating zone technique. The composition of these crystals are: (a) 2.6 $^m/_o$, (b) 4.2 $^m/_o$, and (c) 5.3 $^m/_o$ Y_2O_3. (After Yamakawa, T., Ishizawa, N., Vematsu, K., Mizutani, N., and Kato, M., *J. Cryst. Growth*, 75, 623, 1986.)

A. Microstructure

The precipitate microstructure of single crystals is virtually identical to that observed in polycrystalline materials.[40] The precipitate size also varies slightly with fabrication technique, cooling rate, and composition. Scanning electron microscope (SEM) observations of the precipitates of polished and hot H_3PO_4 etched samples of Y-PSZ single crystals containing 1 to 10 $^m/_o$ Y_2O_3 prepared from float zone specimens[42] are shown in Figure 32. These authors have also determined the volume fractions of m- plus t-ZrO_2 and precipitate content with Y_2O_3 addition (Figure 33). The volume fraction of t-ZrO_2 has been determined by Ingel,[40] (Figure 34) and is compared with prediction based upon assumptions of the location of the cubic and cubic-tetragonal phase boundary between 1700 and 1900°C. Ingel[40] also measured

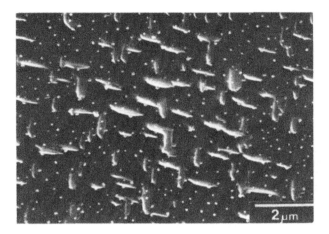

FIGURE 32c.

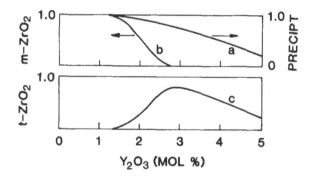

FIGURE 33.   Volume fractions of (a) precipitates, (b) monoclinic,
and (c) tetragonal content of Y-PSZ single crystals with yttria content. (After Yamakawa, T., Ishizawa, N., Vematsu, K., Mizutani,
N., and Kato, M., *J. Cryst. Growth,* 75, 623, 1986.)

the aspect ratio of the t-ZrO_2 precipitates with Y_2O_3 content under presumably identical
cooling conditions (Figure 35). The density and cubic lattice parameters of these crystals
were also measured[40,42] and are compared with a vacancy model by Aleksandrov et al.[45]
(Figure 36) and lattice parameter measurements of Scott[2] and Aleksandrov et al.[45] (Figure
37). The microstructure of the 3 $^m/_o$ Y-PSZ single crystals rapidly cooled (>400°C/hr) by
Michel et al.[47,48] was substantially different from that reported by Ingel[40] or Mizutani and
Kato.[42,46] During the cooling process, the crystals underwent the c → t phase transition
which induced a domain microstructure. Michel et al.[47,48] reported on PSZ crystals stabilized
with Yb, Y, and Gd oxides that contained only a t-ZrO_2 phase, i.e., were free of monoclinic
and cubic phases. Darkfield transmission electron microscope (TEM) observations reveal
the microdomain structure shown in Figure 38. The mean size of the domains is about 50
nm with antiphase boundaries between the domains. The domains have an elongated morphology with {100} habit planes.

B. Mechanical Properties

The hardness, modulus, strength, and fracture toughness of Y-PSZ single crystals have
been evaluated by Ingel[40] and Mizutani and Kato.[46] The results, where possible, are compared

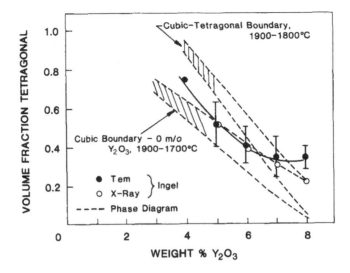

FIGURE 34.   Volume fraction of tetragonal zirconia content of Y-PSZ
single crystals prepared by the skull-melting technique and compared with
predictions based on the $ZO_2 - Y_2O_3$ phase diagram. (After Ingel, R. P.,
Ph.D. thesis, 1982.)

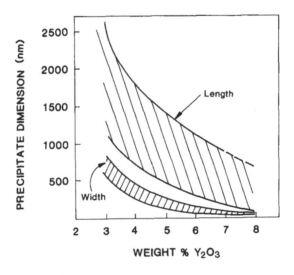

FIGURE 35.   Precipitate dimensions, length, and width for Y-PSZ single
crystals fabricated using the skull-melting technique. There is a much
greater variability in precipitate length than width. (After Ingel, R. P.,
Ph.D. thesis, 1982.)

between the skull melted and floating zone studies. Variation of the hardness with composition is shown in Figure 39. Very similar observations of hardness have been observed
for arc-melted Y-PSZ compositions by Sakuma et al.[52] Ingel[40] found that well-developed
slip bands formed about the indentation site similar to those observed in other polycrystalline
PSZ specimens.[51,52] The hardness maximum occurs for compositions with minimal transformable t-ZrO_2 present. This observation is in agreement with that of Swain and Hannink[33]
who showed the monoclinic-ZrO_2 phase has a much lower hardness. Cubic and tetragonal
phases have about the same hardness, 13.5 GPa, while monoclinic has 7.3 GPa.

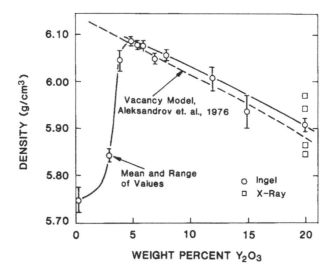

FIGURE 36. Measured density of Y-PSZ single crystals compared with a vacancy model proposed by Aleksandrov et al.[45] The departure from prediction at low Y_2O_3 content occurs because of the presence of monoclinic zirconia. (After Ingel, R. P., Ph.D. thesis, 1982.)

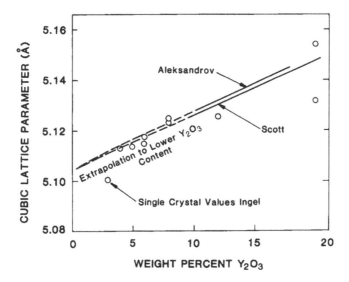

FIGURE 37. Comparison of the single crystal cubic lattice parameter with Y_2O_3 content with reported literature values. (After Ingel, R. P., Ph.D. thesis, 1982.)

The elastic modulus variation with composition has been determined by Mizutani and Kato[46] and found to be strongly dependent upon composition (Figure 40). A minima occurs at approximately 4 mol % Y_2O_3. Ingel[40] measured the variation of Young's Modulus with orientation in the (110) plane for c-ZrO_2 single crystals (Figure 41).

The flexural strength of relatively small samples, $2 \times 3 \times 15$ mm specimens broken in three-point bending on a 12 mm span, show a very pronounced maximum at 2.5 to 3 mol % Y_2O_3.[40] Comparison of float zone, skull melted, and polycrystalline Y-TZP results are shown in Figure 42. Ingel[40] performed all his tests on ground specimens and reported strengths

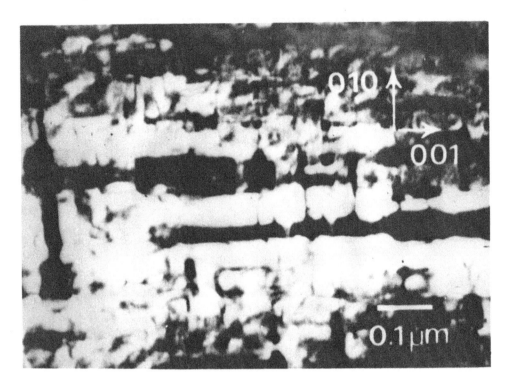

FIGURE 38. TEM observation of the domain microstructure of a tetragonal zirconia single crystal containing 3 mol % Y_2O_3 prepared by rapid cooling. Dark field imaging (g = 110) of an oriented platelet were used. The indexing of the domains is relative to the fluorite cubic lattice. (After Michel, D., Mazerolles, L., and Perez y Jorba, M., *Advances in Ceramics,* Vol. 12, American Ceramic Society, Columbus, Ohio, 1984, 131.)

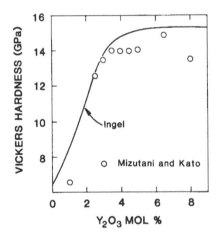

FIGURE 39. Dependence of Vickers hardness with Y_2O_3 content for skull melts and floating zone single crystals.

approaching 2 GPa. The preparation details on the float zone specimens are unknown, but the similarity of the results between the two preparation methods is very striking. SEM observations of the fracture origins by Ingel showed that crack initiation was from grinding flaws. The decrease in strength at lower Y_2O_3 concentrations corresponds to the presence of m-ZrO_2 in the material. However, more detailed examination of both sets of data reveal

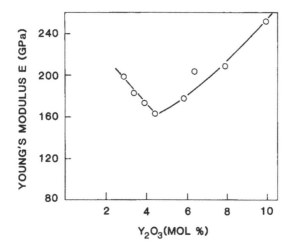

FIGURE 40.   Dependence of Young's Modulus of Y-PSZ single crystals on Y_2O_3 content. (After Mizutani, N. and Kato, M., *Uechida Rokakuho,* 2, 27, 1984.)

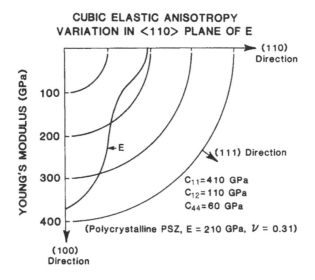

FIGURE 41.   Variation of Young's Modulus with orientation in the {110} plane for fully stabilized cubic zirconia, 10 mol % Y_2O_3. (After Ingel, R. P., Ph.D. thesis, 1982.)

that materials just to the left of the peak strength were still predominantly cubic plus tetragonal ZrO_2. An alternative proposal for the onset of the reduction of strength, namely the stress-induced t to m-ZrO_2 transformation with ensuing microcracks and R-curve behavior, has been proposed in Chapter 3, Section III.E.

The variation of strength with temperature of 6 and 20 wt % Y_2O_3 single crystals is compared in Figure 43. The strength of CSZ (20 wt % Y_2O_3) is almost independent of temperature until 1500°C, whereas the PSZ material decreases rapidly from room temperature to 500°C before attaining temperature independence. At 1500°C, the strength of CSZ is strongly stress-rate sensitive with considerable ductility, and 7% strains are observed at a strain rate of 3.0×10^{-5} per second. The PSZ material, however, shows minimal plastic

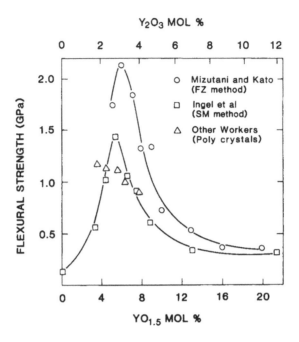

FIGURE 42. Flexural strength of Y-PSZ single crystals prepared by the floating zone[46] and skull melting technique.[40] Also shown are polycrystalline values of strength.

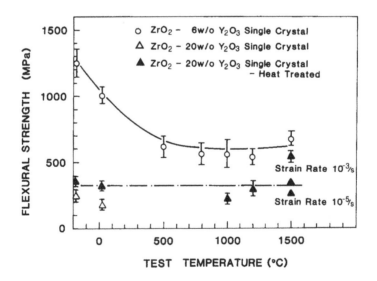

FIGURE 43. Temperature dependence of the flexural strength of partially (6 wt % Y_2O_3) and fully stabilized (20 wt % Y_2O_3) single crystals. (After Ingel, R. P., Ph.D. thesis, 1982.)

strain at this temperature as the t-ZrO_2 precipitates limit dislocation mobility. The rapid decrease in strength with temperature of the PSZ material is related to the increased stability of the t-ZrO_2 as discussed earlier. Similar decline in strength is observed for all PSZ materials, although the base strength of most coarser-grained materials is much lower than observed for PSZ single crystals.

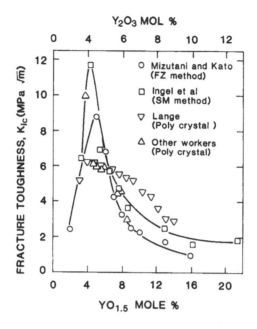

FIGURE 44. Vicker's indentation-derived fracture toughness of Y-PSZ single crystals with Y_2O_3 content prepared by the floating zone (FZ) and skull melting (SM) technique. Also shown are some polycrystalline Y-PSZ (TZP) results using the same technique.

Fracture toughness, as measured by the Vickers indentation method with composition, is shown in Figure 44, which again compares materials of different origins. For both the float zone and skull-melt crystals, the peak in toughness does not coincide with the peak in strength. Ingel[40] found that a toughness maxima occurred at 2 $^m/_o$ Y_2O_3, whereas the peak strength was measured at 2.5 $^m/_o$. Mizutani and Kato[46] found that peak strength was at 3 $^m/_o$ and peak toughness at 2.5 $^m/_o$. The values of peak toughness did occur at the highest concentration of t-ZrO_2. Ingel[40] observed a slight difference between double cantilever beam (DCB) and indentation-determined K_{IC} with the indentation giving significantly higher values at peak toughness (12 MPa$\sqrt{m}$ versus 8 MPa$\sqrt{m}$). For cubic-stabilized-zirconia (CSZ) materials, there was virtually no difference between the two measurements. The rapid fall off in K_{IC} with increasing Y_2O_3 beyond the maximum reflects the decrease in t-ZrO_2 content and precipitate size. It is clear there are additional toughening mechanisms present in these materials, for example, the comparison of the temperature dependence of K_{IC} of partially (3 $^m/_o$) and fully stabilized (10 $^m/_o$) single crystals (Figure 45). These measurements, made on small single-edge notched-bend (SENB) specimens and at high temperature (>1300°C), reflect ductility at the notch root prior to fracture. Note at 1000°C there is no transformation toughening, yet the toughness is double that of the matrix. This increase is of the order of the increase anticipated for crack deflection by rods dispersed in the matrix.[54]

The toughness of single phase t-ZrO_2 crystals, formed by rapid cooling of skull-melted materials, is comparable to those reported above. Michel et al.,[47,48] using the indentation method, measured a K_{IC} of 6 MPa$\sqrt{m}$. There was no evidence from Raman spectroscopy, TEM, or XRD of any transformation adjacent to the fracture surface. However, these authors noted a difference in fracture morphology between single-phase cubic and tetragonal-ZrO_2 crystals. The cubic crystals indented on the (001) surface preferred to cleave along (111) planes, whereas the tetragonal crystals fractured along (100) planes. The fracture surfaces associated with the tetragonal crystals were very rough (Figure 46), with microfacets typically

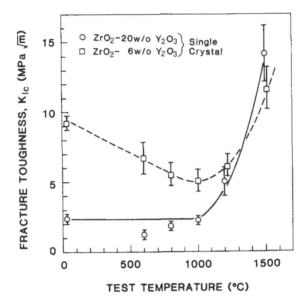

FIGURE 45.   Fracture toughness measurements using the single edge notched bend (SENB) technique for a partially (6 wt % Y_2O_3) and fully stabilized (20 wt % Y_2O_3) Y-PSZ single crystals as a function of temperature. (After Ingel, R. P., Ph.D. thesis, 1982.)

50 to 100 nm formed by fracture along the three {100} cubic planes. It is considered that the increased toughness is associated with the severe crack deflections induced by the domain structure of these materials. Michel et al.[47,48] propose that fracture along well-defined crystallographic planes can be explained in two ways: the internal stress fields associated with the t-lattice distortion which can induce the crack deflection at domain boundaries, or preferential cleavage along {100} planes.

There has recently been some questioning of the explanation of the high toughness in these Y-PSZ single crystal materials. Virkar and Matsumoto[55] suggest that reversible-phase transformations may be occuring on the basis of ferro-elastic behavior. The high toughness experienced and the virtual absence of transformation on the fracture surfaces are further supported by the observations of the slip band like features about Vickers indents by Ingel[40] on materials with 3 to 5 wt % Y_2O_3. This behavior has been observed by Virkar and Matsumoto[55] for other TZP materials. Evans and Cannon[56] have proposed that the toughening associated with this mechanism is proportional to the magnitude of the energy loss in the ferro-elastic hysteresis loop.

A comparison of the properties of other single crystals of Mg-PSZ and Ca-PSZ, as well as some data for polycrystalline material, are listed in Table 3. The composition and properties of these materials were not optimized.

The thermal expansion of a range of PSZ single-crystal materials has been measured by Larsen and Adams.[57] Materials containing more than 5 $^w/_o$ Y_2O_3 displayed linear behavior upon heating and cooling from 1500°C; the data is tabulated in Table 4. They noted that materials containing less Y_2O_3 displayed hysteresis with often significant differences between the first and second heating cycle. The Ca- and Mg-PSZ specimens exhibited hysteresis upon cooling, presumably because of precipitate coarsening at 1500°C during the heating cycle and destabilization of t-ZrO_2 upon cooling.

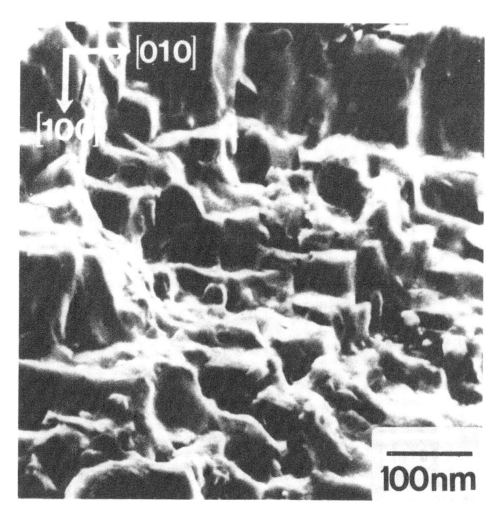

FIGURE 46.   SEM of a fracture surface of a rapidly cooled Y-PSZ single crystal containing 3 mol % of Y_2O_3. The micrograph illustrates the influence of the tetragonal domains on the fracture path. The indexing is relative to the fluorite cubic lattice. (After Michel, D., Mazerolles, L., and Perez y Jorba, M., *Advances in Ceramics*, Vol. 12, American Ceramic Society, Columbus, Ohio, 1984, 131.)

VIII. Y-TZP

TZP are, by definition, monophase (t-ZrO_2), but since the composition range and temperature range for fabrication of such are limited, this restriction must be broadened. Here we shall consider TZP materials, as defined in the introduction, as fine-grained PSZ (typically <1 μm) with compositions varying from 1.75 to 3.5 mol % Y_2O_3. The first description of such materials with attractive mechanical properties was by Gupta et al.[58,59] Since then many reasearch groups have fabricated such materials and powders to produce these materials. They are now available in semi-production quantities by a number of companies. The powders and sintering conditions were discussed earlier in Section II.

A. Microstructure

The microstructure and composition of a number of TZP materials from various manufacturers have been compared by Ruhle et al.[60] They found a wide range in compositions (3.5 to 8.7 wt % Y_2O_3) with t-ZrO_2 content varying from 100 to 60%, the remaining phase being c-ZrO_2. In all instances they found a glassy-grain-boundary phase rich in silica and

Table 3

COMPARISON OF PHYSICAL AND MECHANICAL PROPERTIES OF ZrO₂ ALLOYS

Single crystal	Composition (wt %)	Precipitates (vol %)	Density (kg/m³)	Hardness (GPa)	Young's modulus (GPa)	Flexural strength (MPa)	Fracture toughness (MPa·m$^{1/2}$)
Mg-PSZ	2.8	48	5790	14.4	200	685 ± 48	4.82 ± 0.56
Ca-PSZ	4.0	38	5850	17.1	210	661 ± 33	3.97 ± 0.46
Ca-CSZ	9	0	5680	17.2	210	241 ± 28	2.54 ± 0.13
Y-PSZ	5	52	6080	13.6	233	1384 ± 80	6.92 ± 0.14
Y-CSZ	20	0	5910	16.1	233	346 ± 58	1.91 ± 0.16

Polycrystalline zirconia	Young's modulus (GPa)	Flexural strength (MPa)	Fracture toughness (MPa·m$^{1/2}$)
Mg-PSZ	200	430—700	4.7—15
Ca-PSZ	200—217	400—650	5.0—9.6
Y-PSZ	210—238	696—980	5.8—9.0

From Ingel, R. P., Lewis, D., Bender, B. A., and Rice, R. W., *Advances in Ceramics*, Vol. 12, Science and Technology of Zirconia II, Claussen, N., Ruhle, M., and Heuer, A. H., American Ceramic Society, Columbus, Ohio, 1984, 408. With permission.

Table 4

THERMAL EXPANSION OF SINGLE-CRYSTAL ZIRCONIA AT 1000°C[a]

Additive	Percent linear expansion
No stabilizer	0.8[b]
2.8% MgO	0.2[c]
4% CaO	0.92[c]
3% Y₂O₃	0.035[c]
4% Y₂O₃	0.99[c]
5% Y₂O₃	1.02[d]
6% Y₂O₃	1.00[d]
7% Y₂O₃	0.96[d]
8% Y₂O₃	1.03[d]
12% Y₂O₃	0.92[d]
20% Y₂O₃	0.96[d]

[a] Upon first heating cycle.
[b] Appears unstabilized from thermal expansion behavior.
[c] Appears partially stabilized from thermal expansion behavior.
[d] Appears fully stabilized from thermal expansion behavior.

yttria varying in width from 2 to 100 nm. Grain sizes varied depending upon sintering time, temperature, and composition. The t-ZrO₂ were generally <1 μm, whereas the c-ZrO₂ grains were much larger. Examples of the composition dependence of the grain size for one manufacturer (Toyo Soda) sintered at 1500°C for 1 hr[61] are shown in Figure 47. The larger grains in the 4 and 6 mol % compositions are c-ZrO₂ grains. The approximate phase composition of these materials is shown in Figure 48 which agrees tolerably well with the Y₂O₃-ZrO₂ diagram of Scott.[2] Comparison between the t + c and c boundaries determined

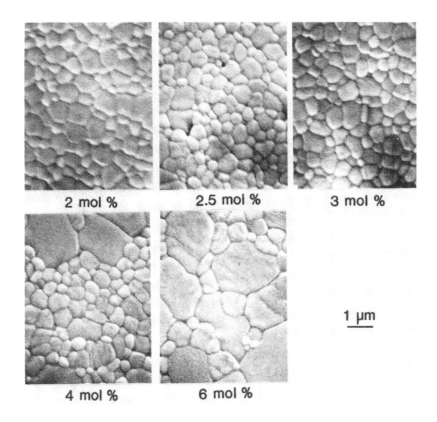

FIGURE 47. SEM of thermally etched Y-TZP materials of varying compositions from 2 to 6 mol % Y_2O_3. All specimens had been sintered for 1 hr at 1500°C. The bar length is 1μm in all the micrographs. (After Gross, V. and Swain, M. V., *J. Aust. Ceram. Soc.*, 22, 1, 1986.)

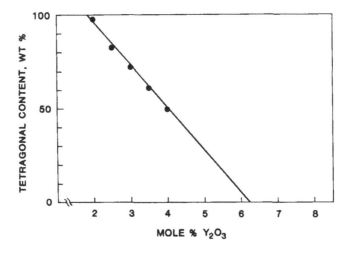

FIGURE 48. Tetragonal phase content of Y-TZP materials sintered at 1500°C with different mol % Y_2O_3.

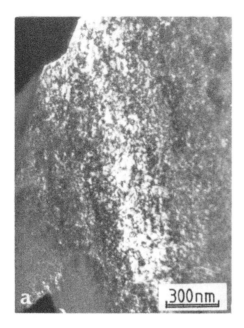 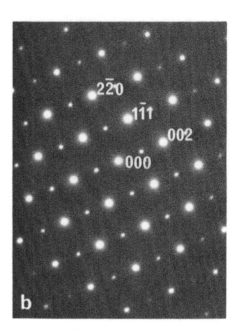

FIGURE 49.   (a) Transmission electron bright field image of tetragonal precipitates in cubic zirconia grain of a Y-TZP material. (b) Corresponding electron diffraction pattern from (a) revealing both intense c-ZrO$_2$ and weaker t-ZrO$_2$ diffraction spots. Micrograph courtesy of J. Drennan, CSIRO.

by detailed energy-dispersive analysis (TEM) and the XRD analysis were shown in Chapter 2, Figure 21.

The TEM observations revealed that the c-ZrO$_2$ particles contained small t-ZrO$_2$ precipitates. This evidence is shown clearly in Figure 49 which is a bright field image of a c-ZrO$_2$ grain and corresponding diffraction pattern revealing both intense c-ZrO$_2$ and weaker t-ZrO$_2$ diffraction spots. The t-ZrO$_2$ precipitates are ~20 nm in size.

B. Mechanical Properties

Optimal values of the flexural strength of sintered and sintered plus hot isostatic pressing (HIPing) of TZP powders are shown in Figure 50. Maximum strengths have been observed at approximately 3 mol % Y$_2$O$_3$ with a sharp decrease at higher Y$_2$O$_3$ contents and below 2 mol % Y$_2$O$_3$.[62,63] The increase in strength with HIP is associated with an improvement in density and reduction, but not complete elimination of flaws.

Variation in toughness of Y-TZP materials with composition is shown in Figure 51. Tsukuma et al.[64] find that K$_{IC}$ is independent of sintering temperature at the various compositions. They also find that the monoclinic content of the fracture surface decreases with decreasing K$_{IC}$ and increasing yttria. It is possible to use the fracture surface m-ZrO$_2$ to determine the transformed zone depth.[17,18]

The fracture toughness of Y-TZP materials is related to the size of the t-ZrO$_2$ grains. The critical size of such grains is still an area of controversy and is discussed in Chapter 3. It is possible to vary the grain size of one composition by isothermal heat treatment, thereby ensuring the phase content remains constant. Figure 52 is a plot of the grain size dependence for 2 mol % Y-TZP held at 1500°C, of (a) K$_{IC}$, (b) fracture surface m-ZrO$_2$ content, and (c) transformation zone depth.[65] These observations show the strong dependence of toughness on grain size which is reflected in the fracture surface monoclinic content and transformation zone depths.

The metastability and high toughness associated with larger grain-sized Y-TZP materials

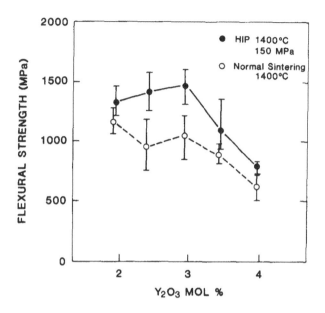

FIGURE 50. Flexural strength dependence on Y_2O_3 content of sintered and hot isostatically pressed (HIP) Y-TZP materials. Sintering was carried out at 1350°C for 2 hr prior to HIP. (After Tsukuma, K., Kubota, Y., and Tsukidate, T., *Advances in Ceramics*, Vol. 12, American Ceramic Society, Columbus, Ohio, 382, 1984.)

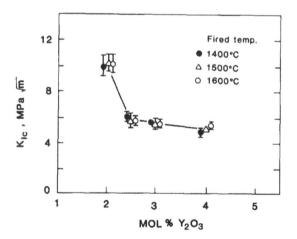

FIGURE 51. Fracture toughness measured using the Vicker's indentation technique for Y-TZP materials as a function of Y_2O_3 content and sintering temperature. (After Tsukuma, K., Kubota, Y., and Tsukidate, T., in *Advances in Ceramics*, Vol. 12, American Ceramic Society, Columbus, Ohio, 382, 1984.)

is offset by their instability under moist environments. Figure 53 shows (a) the variation of surface $m\text{-}ZrO_2$ content, and (b) flexural strength for 3 mol % Y-TZP as a function of grain size and aging time at 230°C in air. Only materials with grain sizes less than 0.4 μm show minimal degradation. However, as shown by Sato and Shimada,[66] the effect of hot water on the destabilization is much more severe than warm moist air. Figure 54 compares the surface $m\text{-}ZrO_2$ content of 3 mol % Y-TZP sintered at 1600°C (1.1 μm grain size) and held

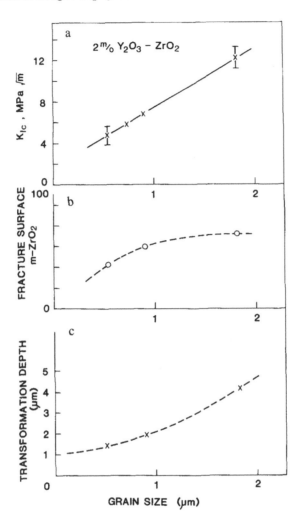

FIGURE 52. Grain size dependence of (a) the fracture toughness (K_{IC}), (b) monoclinic generated on the fracture surface, and (c) depth of the transformed zone determined from (b) for 2 mol % Y-TZP. The grain size was coarsened by isothermal aging at 1500°C. (After Swain, M. V., *J. Mater. Sci. Lett.*, 5, 1159, 1986.)

for comparable times in air and water at various temperatures. These same authors found that the rate of transformation was directly proportional to the concentration of t-ZrO_2 on the surface, and that nonaqueous solvents with low pair electron orbitals opposite a proton donor site also greatly enhanced the tranformation. They suggested that a reaction system similar to that proposed by Michalske and Freiman[67] for the stress corrosion in vitreous silica was responsible for the decomposition by a crack-growth mechanism.

More recent contributions on the question of stability of Y-TZP materials have suggested that other factors are important. Schmauder and Schubert[68] have proposed that internal stresses resulting from thermal expansion anisotropy (TEA) of the tetragonal grains may play a significant role. They find that the magnitude of the TEA stresses increases with decreasing yttria content in accordance with the change in observed and predicted tetragonal particle stability. Another factor that these same authors, and more recently Lange et al.[69] have observed, is the variation in Y_2O_3 content from grain to grain in the original powders

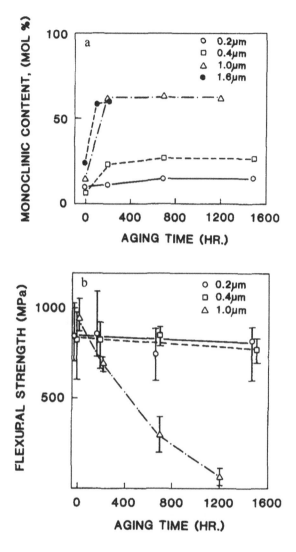

FIGURE 53. Relationship between (a) monoclinic zirconia content and (b) flexural strength of 3 mol % Y-TZP materials of differing grain size with aging time at 230°C. The rapid onset of monoclinic on the tensile surface is associated with a significant reduction in strength. (After Tsukuma, K., Kubota, Y., and Tsukidate, T., in *Advances in Ceramics*, Vol. 12, American Ceramic Society, Columbus, Ohio, 382, 1984.)

and sintered grains. The basic conclusions of the significance of TEA stresses is supported by observations by Masaki[70] who reports on the importance of composition, grain size, and final density of Y-TZP materials. The higher the density, the greater the resistance to moisture-assisted strength degradation. The presence of pores, coupled with residual TEA stresses, provides the stress concentration sites for moisture-assisted destabilization to take place. Hot isostatic pressing plays an important role in that it ensures the achievement of high densities with minimal grain growth and, hence, the greatest resistance to strength degradation in service.

An alternative explanation of the strength degradation has been proposed by Nakanishi et al.[71] These authors claim that the strength degradation is due to microcrack propagation induced by the expansion of isothermally produced m-ZrO$_2$. They found that a T-T-T curve

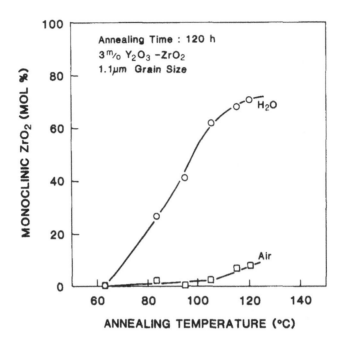

FIGURE 54. Development of monoclinic zirconia on the surface of polished specimens of 3 mol % Y-TZP materials after annealing for 120 hr at various temperatures in water and air. Specimens were sintered at 1600°C and had a mean grain size of 1.1 μm. (After Sato, T. and Shimada, M., *J. Am. Ceram. Soc.*, 68, 356, 1985.)

could be constructed from isothermal aging studies in the temperature regime 200 to 350°C. The nose of this curve was dependent upon the grain size, composition, and character of the grain boundaries.

Various approaches have been adopted to increase the stability of Y-TZP materials including limiting initial grain size and grain-boundary phases, silicate phases, coating the surface with a thin, more stable c-ZrO_2 phase,[72] and addition of Al_2O_3 to Y-TZP (see Chapter 5).

IX. Ce-TZP

There have been limited studies of this system, although recent work by Tsukuma and Shimada[73] and Tsukuma[74] have been thorough. Lange[75] investigated Ce-PSZ-Al_2O_3 composite systems. Coyle et al.[76] have also considered this area, but publications are limited to a thesis by Coyle.[77]

Ce-TZP materials displayed considerably better thermal stability than Y-TZP under comparable cycling and hydrothermal conditions.[73] Only materials containing <10 $^m/_o$ CeO_2 (impure CeO_2) developed m-ZrO_2 at the surface upon thermal cycling to 500°C for 1000 hr. Hydrothermal aging at 150°C for 500 hr did not influence the mechanical properties or phase composition of materials with 12 mol % CeO_2 (pure) and 2.5 μm grain size. Y-TZP materials containing 2 and 3 mol % Y_2O_3 and 0.2 μm grain size degraded substantially under similar conditions (Figures 55 and 56). The increased stability of Ce-TZP is not properly understood.

The CeO_2-ZrO_2 phase diagram has been presented in Chapter 2, Figure 22 and shows a wide region where stabilization of t-ZrO_2 is possible. Lange[75] reported that compositions with <12 $^m/_o$ CeO_2 contained m-ZrO_2. Tsukuma[74] found that the critical composition for

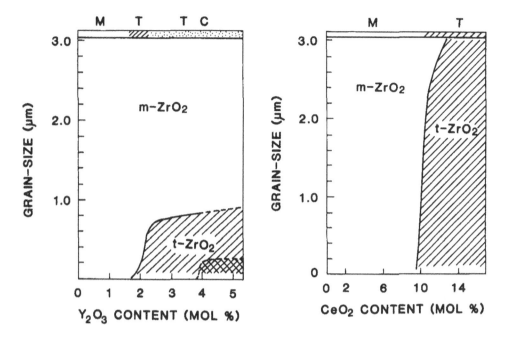

FIGURE 55. Phase-composition grain-size stability diagram for Y-TZP materials exposed to moist air (hatched) and water (crosshatched) at 150°C. (After Tsukuma, K., *Am. Ceram. Soc. Bull.*, 65, 1386, 1986.)

FIGURE 56. Composition-grain size-phase stability diagram for Ce-TZP materials exposed to moist air at 150°C. (After Tsukuma, K., *Am. Ceram. Soc. Bull.*, 65, 1386, 1986.)

spontaneous transformation was dependent upon sintering temperature and grain size. The boundary of the grain size dependence of t-ZrO_2 vs. CeO_2 content of the phase diagram is shown in Figure 56 and can be compared with the Y_2O_3-ZrO_2 diagram (Figure 55). Tsukuma and Shimada[73] found that fabrication of CeO_2-ZrO_2 materials from commercial grade $CeCl_3$ which contained 15% ($LaCl_3$ and $NdCl_3$) led to greater stability of the t-ZrO_2 phase as well as the presence of minor pyrochlore compound.

These materials across the composition range 12 to 20 $^m/_o$ CeO_2 were single-phase TZP, whereas in the Y-TZP materials, this only occurs over a very narrow range (2 to 3 $^m/_o$ Y_2O_3) with c-ZrO_2 present at higher concentrations. Hot pressing or HIPing of Ce-ZrO_2 materials is limited since CeO_2 readily reduces to Ce_2O_3 in the presence of carbon-containing atmospheres (graphite dies or heating elements). Grain size of Ce-ZrO_2 materials is somewhat larger than Y-TZP bodies sintered under similar conditions.

Tsukuma and Shimada[73,74] measured the mechanical properties of materials prepared from pure and commercial CeO_2 sources. The results are more exhaustive with the commercial CeO_2 (containing La and Nd impurities), but were virtually identical for both. The Vickers hardness H_V, three-point bend strength (30 mm span) and fracture toughness (measured from Vickers indentation tests (~490 N) are plotted in Figure 57. The results are plotted for two grain sizes, the coarser material having lower hardness and strength, but higher toughness particularly at lower CeO_2 content. A DCB evaluation of toughness for the 1 μm grain size material suggests the indentation technique significantly overestimated K_{IC} for these materials (see Chapter 6, Section I.B). The stress-strain curves of the 12 $^m/_o$ material were considerably nonlinear before fracture, with the onset of this nonlinearity decreasing with increasing grain size (Figure 58). Optical observations of the tensile surface of a 12 $^m/_o$ Ce-TZP material showed the development of Luders-like bands when the elastic limit was exceeded. Examples of these observations are shown in Figure 59 and they consist of localized zones of transformation about 4 to 5 grain size that, in this instance, initiated at one edge and slowly

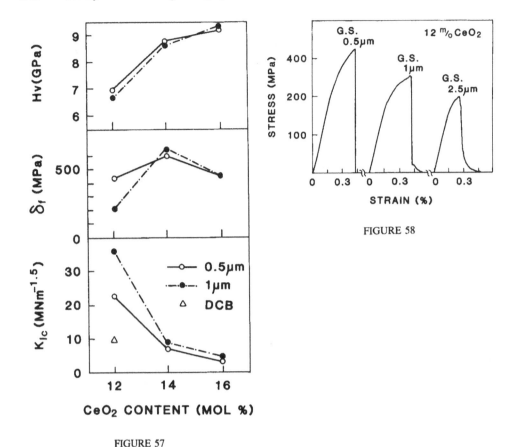

FIGURE 57

FIGURE 57. Dependence of mechanical properties of Ce-TZP materials on composition and grain size (a) Vickers hardness, (b) flexural strength, and (c) fracture toughness. (Tsukuma, K., *Am. Ceram. Soc. Bull.*, 65, 1386, 1986.)

FIGURE 58. Stress strain curves of Ce-TZP materials containing 12 mol % CeO_2 as a function of grain size. (After Tsukuma, K., *Am. Ceram. Soc. Bull.*, 65, 1386, 1986.)

spread across the material. There was also little evidence for cracking about Vickers indentations in the same material even at loads of 50 kg. This was presumably because of the large transformed zone about the impression suggested by the much lower hardness and the difficulty of crack initiation under these conditions (see Chapter 6, Section I.B). The values of toughness measured in this way may be overestimates for the lower ceria containing Ce-TZP materials. Similar problems would arise using the modified indentation technique of Cook and Lawn.[78]

An example of the deformation about an 80 kg Vickers indentation in a 12 $^m/_o$ Ce-TZP material with 2.5 μm grain size is shown in Figure 60. There is very little evidence for microcracking, but substantial stress-induced transformation manifested by the extensive uplift from shear-like bands emanating from about the impression.[79] It is not possilbe to determine a K_{IC} value for this material from the observation in Figure 60. Independent determination of the K_{IC} using a DCB geometry[80] gave a steady state value of only 9 MPa$\sqrt{m}$ with an extensive R-curve of ~ 1 mm and a transformed zone up to 20 to 30 μm either side of the crack and up to 200 μm ahead of the crack tip.

TEM observations of the deformation about a Vickers indent confirms that the material is substantially transformed, but the extent of transformation decreases with distance from the indent. Examples of the transformation at locations marked in Figure 60 are shown in Figure 61. Adjacent to the indent in Figure 61a, the individual grains are fully transformed

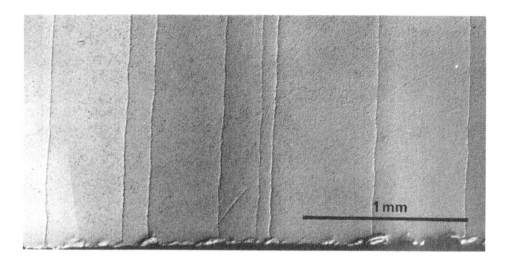

FIGURE 59. Optical micrographs using interference contrast of Luders-like bands of transformed grains on the tensile surface of a 12 mol % Ce-TZP material loaded in flexure.

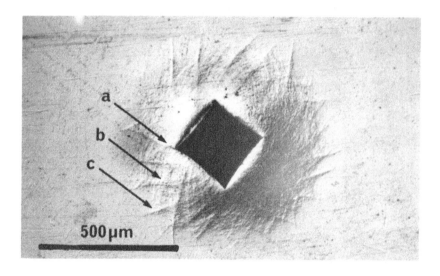

FIGURE 60. Optical micrograph of the deformation surrounding on 80 kg indentation in a 12 mol % Ce-TZP material showing the almost complete absence of radial cracking. Areas indicated as a, b, and c are shown in Figure 61.

with a number of intersecting m-ZrO$_2$ variants in an individual grain. Whereas near the edge of the zone in a shear-like band only partial transformation of a grain has taken place Figure 61c.

TEM observations of partially transformed grains at the edge of deformed regions, induced by indentation, or uni- or biaxial stress loading, suggests that Ce-TZP materials (and 2 $^m/_o$ Y-TZP) accommodate large shear strains by a mechanism which may be termed "transformation-induced plasticity".[79] This process is achieved by the preferential nucleation (often autocatalytic) and propagation of a m-variant which will counteract the overall shear strain within a grain. The propagation and the extent of transformation within the grain will be a

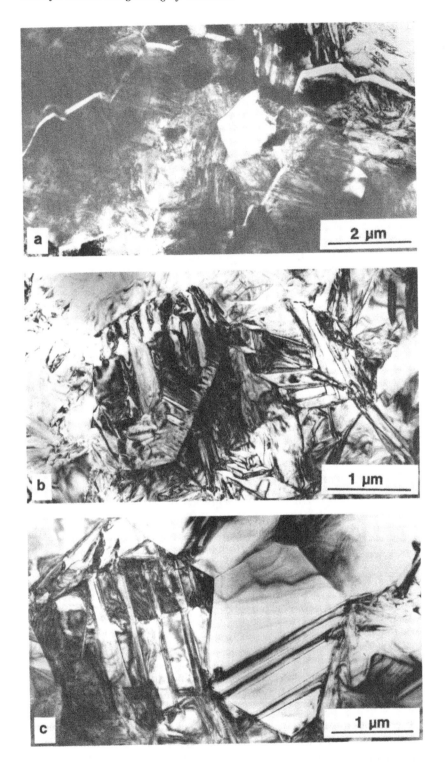

FIGURE 61.   Bright field TEM image of the transformation about a Vickers indent, similar to that shown in Figure 60 at positions marked a, b, and c. Adjacent to the indentation, (a) the grains are fully transformed and show intergranular fracture. Transformation "intensity" decreases with distance from the indent (b), such that at the edge of the deformed zone, at (c), only partial transformation of the grains takes place.

function of the applied stress.* As the stresses imposed upon the grain change, due in part by transformation of the surrounding grains, a new m variant is nucleated and propagates to counteract the new stress system. This situation is replicated until the whole grain has been transformed. When this point has been reached, and the applied stress is still increasing, a situation similar to the fully work-hardened state in metals can be envisaged and the ceramic will only accommodate additional stress by fracture.

The beginnings of the above scenario are depicted in Figure 62. Figure 62a shows a grain containing five laths of the same m-variant. The selected area electron diffraction pattern of this grain (Figure 62b), shows the grain matrix orientation as $[001]_t$ (intense spots) with the m laths in the $[020]_m$ orientation (less-intense spots). This situation is summarized schematically in Figure 63. The partial transformation of grains and the ease of autocatalytic nucleation may, in part, account for the low K_{IC} (9 MPa$\sqrt{m}$) in the presence of the large transformation zone and extensive R-curve behavior.

Another feature of the t $\rightarrow$ m transformation in 12 $^m/_o$ % Ce-TZP materials is its readily reversible nature. From experiments on transformations in sections, A_s-A_f is at about 100°C. Such low temperatures are readily achieved on grinding and at fracture surfaces. Therefore, the reversible nature of the t $\rightarrow$ m transformation implied that Ce TZP materials may be able to sustain repeated plastic deformation via the transformation-induced plasticity mechanism.

Experiments carried out to study the reversible nature also indicate strong preferred orientation effects if the transformation in *both* directions (Figure 64) shows the X-ray profile of the $\{111\}_m$ and $\{111\}_t$ and $\{200\}_t$ before and after the transformation. The ratio of the various peaks should be 4:3 for the $11\bar{1}_m$:111_m and 1:2 for the 002:$200/020_t$.

Figure 64 compares the XRD trace of 12 mol % Ce-TZP that has been (a) polished, (b) ground, (c) a polished surface cooled to liquid N_2 (below the M_s temperature $\sim$ -125°C), and (d) a fracture surface. The former materials contain primarily t-ZrO_2 with minor amounts of m-ZrO_2 on the ground surface. As can be seen in Figure 64a of the polished surface, the anticipated ratios of the t-ZrO_2 variants are observed. However, for the ground surface, this ratio has been effectively reversed. For the fracture surface and the material exposed below M_s, the amount of m-ZrO_2 is significantly larger than the ground surface. On the fracture surface (Figure 64d), considerable preference for the formation of the (111) m-ZrO_2 is observed.

X-ray pole figure measurements of ground surfaces, to determine whether texturing of the t-ZrO_2 variants has occurred due to grinding, suggested this not to be the case. Some types of texturing of this type would have been anticipated had the ferro-elastic switching mechanism proposed by Virkar and Matsumoto[55] occurred. An alternative explanation of the observations is that a t $\rightarrow$ m $\rightarrow$ t-ZrO_2 takes place. The mechanical grinding action has two effects, generating high stresses and heat. The former is capable of initiating the t $\rightarrow$ m-ZrO_2 transformation, like a pointed indentation (see Figure 60), whereas the heat generated is responsible for the m $\rightarrow$ t-ZrO_2 reverse transformation. Strong evidence in support of such a proposition was observed by grinding less severely; for example, by slowly grinding in a liquid medium on wet and dry abrasive paper or SiC grit. The X-ray traces are virtually identical with the fracture surface trace (Figure 64c). A more detailed account of these observations is given elsewhere.[81]

Therefore, in the context of the free energy forms, presented schematically in Chapter 2, Figure 6b, the free energy of the m-phase in 12% Ce-TZP is in a state between $F_{mono(S)}$ (small constrained particles) and $F_{mono(C)}$ (critically sized constrained particles), such that the

---

* Partial transformation in Ce-TZP and Y-TZP grains suggests that the m nucleating and propagating stress are about the same. This is in contrast to ZTC, in which the M_s is above room temperature, which will spontaneously transform once a nucleus has been activated.

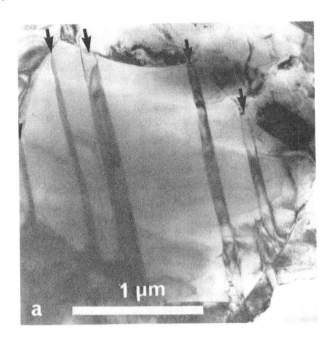

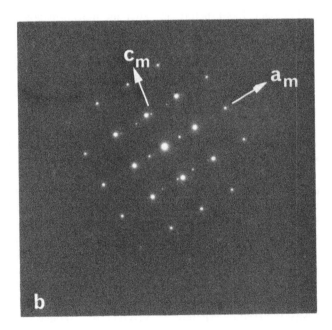

FIGURE 62.   (a) Bright field TEM image of a tetragonal grain containing
five laths of single monoclinic variant (arrowed). (b) Selected area electron
diffraction pattern of the grain, intense spot array $[100]_t$, less intense spot
array from $[010]_m$ oriented monoclinic laths. Pattern not corrected for
rotation.

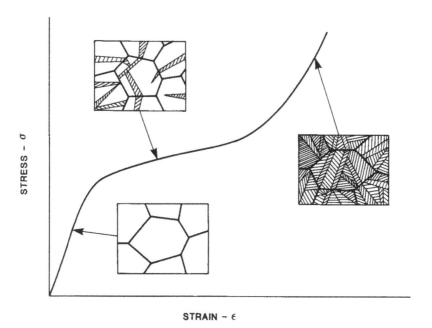

FIGURE 63.   Schematic representation of the transformation plasticity and work hardening of TZP materials.

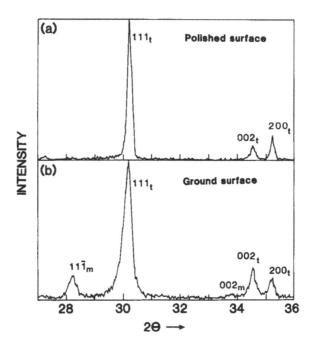

FIGURE 64.   X-ray diffractometer traces of Ce-TZP showing 111 and 200 reflections for tetragonal and monoclinic phases after samples have been (a) carefully polished, (b) ground, using a surface grinder; (c) polished and cooled in liquid nitrogen, and (d) fractured in a double cantilever beam (DCB) test.

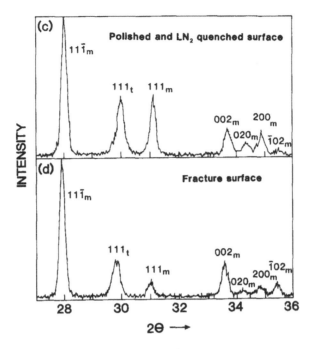

FIGURE 64 (continued)

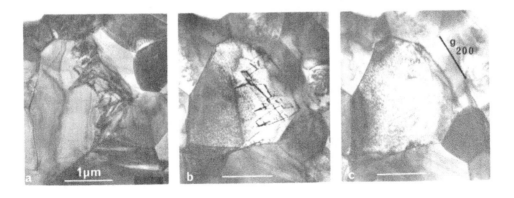

FIGURE 65.   Bright field TEM images of a partially transformed grain in 12 mol % Ce-TZP. (a) Shows the monoclinic laths, (b) the region has now back transformed to tetragonal (the former monoclinic/tetragonal interface shows surface steps), (c) by imaging with a 200_t-type diffraction vector, it can be seen that the whole grain has reverted to a single variant of the tetragonal phase, i.e., no domain boundaries are evident. A similar situation is also observed with fully transformed grains.

stress is not only necessary to nucleate the transformation, but also to propagate and maintain it.

Recent attempts[55] to explain the toughening behavior of ZTC, in terms of ferro-elastic transformations, have not been completely verified. However, observations such as those presented above suggest that the ferro-elastic mechanism is unlikely to be operating, in that the cell axis realigns with the residual stress on back transformation at the surface. Figure 65 shows the completed reverse transformation in a partially transformed grain in the Ce-TZP material. Electron diffraction patterns and images show no evidence for domain structure.

REFERENCES

1. **Miller, R. A., Smialek, J. L., and Garlick, R. G.**, Phase stability in plasma-sprayed, partially stabilized zirconia-yttria, in *Science and Technology of Zirconia*, Advances in Ceramics, Vol. 3, Heuer, A. H. and Hobbs, L. W., Eds., American Ceramic Society, Columbus, Ohio, 1981, 241.
2. **Scott, H. G.**, Phase relationships in the zirconia-yttria system, *J. Mater. Sci.*, 10, 1527, 1975.
3. **Drennan, J. and Hannink, R. H. J.**, Effect of SrO additions on the grain-boundary microstructure and mechanical properties of magnesia-partially stabilized zirconia, *J. Am. Ceram. Soc.*, 69, 541, 1986.
4. **Drennan, J. and Swain, M. V.**, Sintering of magnesia partially stabilized zirconia (Mg-PSZ), Proceedings of Sintering '87, Tokyo, in press.
5. **Swain, M. V.**, Unpublished results.
6. **Martin, J. W. and Doherty, R. D.**, *Stability of Microstructures in Metallic Sytems*, Cambridge Univ. Press, Cambridge, Mass., 1980.
7. **Hannink, R. H. J.**, Growth morphology of the tetragonal phase in partially stabilized zirconia, *J. Mater. Sci.*, 13, 2487, 1978.
8. **Garvie, R. C., Hannink, R. H. J., and Pascoe, R. T.**, U.S. Patent 4,067,745, 1978.
9. **Hannink, R. H. J., Johnston, K. A., Pascoe, R. T., and Garvie, R. C.**, Microstructural changes during isothermal aging of a calcia partially stabilized zirconia alloy, in *Advances in Ceramics*, Vol. 3, American Ceramic Society, Columbus, Ohio, 1981, 116.
10. **Marder, J. M., Mitchell, T. E., and Heuer, A. H.**, Precipitation from cubic ZrO_2 solid solutions, *Acta Metall.*, 31, 387, 1983.
11. **Dickerson, R. M., Swain, M. V., and Heuer, A. H.**, Microstructural evolution of Ca-PSZ and the room temperature instability of tetragonal ZrO_2, *J. Am. Ceram. Soc.*, 70, 214, 1987.
12. **Warlimont, H.**, *Electron Microscopy and Structure of Materials*, Thomas, G., Ed., Univ. of California Berkeley, 1972, 505.
13. **Khatchaturyan, A. G.**, in *Theory of Structural Transformations in Solids*, John Wiley & Sons, New York, 1983.
14. **Lanteri, V., Mitchel, T. E., and Heuer, A. H.**, Morphology of tetragonal precipitates in partially stabilized ZrO_2, *J. Am. Ceram. Soc.*, 69, 564, 1986.
15. **Garvie, R. C., Hannink, R. H. J., and Urbani, C.**, Fracture mechanics study of a transformation toughened zirconia alloy in the $CaO-ZrO_2$ System, Y_2O_3, *Ceramurgica Int.*, 6, 19, 1980.
16. **Swain, M. V., Hannink, R. H. J., and Garvie, R. C.**, The influence of precipitate size and temperature on the fracture toughness of calcia- and magnesia-partially stabilized zirconia, in *Fracture Toughness of Ceramics*, Vol. 6, Bradt, R. C., Hasselman, D. P., and Lange, F. F., Eds., Plenum Press, New York, 1983, 339.
17. **Kosmac, T., Wagner, R., and Claussen, N.**, X-ray determinations of transformation depths in ceramics containing tetragonal ZrO_2, *J. Am. Ceram. Soc.*, 64, C72, 1981.
18. **Garvie, R. C., Hannink, R. H. J., and Swain, M. V.**, X-ray analysis of the transformed zone in partially stabilized zirconia (PSZ), *J. Mater. Sci. Lett.*, 1, 437, 1982.
19. **Hughan, R. R. and Hannink, R. H. J.**, Precipitation during controlled cooling of magnesia-partially stabilized zirconia, *J. Am. Ceram. Soc.*, 65, 556, 1986.
20. **Hannink, R. H. J. and Lowe, B. J.**, Characterization of secondary precipitate growth in magnesia partially stabilized zirconia, *Proc. 12th Australian Ceramics Conf.*, Australian Ceramic Society, Melbourne, Australia, 1986, 155.
21. **Brown, L. M., Coole, R. H., Ham, R. K., and Purdy, G. R.**, Elastic stabilization of arrays of precipitates, *Scr. Metall.*, 7, 815, 1973.
22. **Johnson, W. C.**, On the elastic stabilization of precipitates against coarsening under applied load, *Acta Metall.*, 32, 465, 1984.
23. **Garvie, R. C., Hannink, R. H. J., and McKinnon, N. A.**, U.S. Patent 4,279,655, 1981.
24. **Hannink, R. H. J. and Garvie, R. C.**, Sub-eutectoid aged Mg-PSZ alloy with enhanced thermal up-shock resistance, *J. Mater. Sci.*, 17, 2637, 1982.
25. **Hannink, R. H. J. and Swain, M. V.**, Particle toughening in partially stabilized zirconia: influence of thermal history, in *Tailoring of Multiphase and Composite Ceramics Conference*, Messing, G. L., Pantano, C. G., and Newnham, R. E., Eds., Plenum Press, New York, 1986, 259.
26. **Hannink, R. H. J.**, Microstructural development of sub-eutectoid aged $MgO-ZrO_2$ alloys, *J. Mater. Sci.*, 18, 457, 1983.
27. **Swain, M. V., Garvie, R. C., and Hannink, R. H. J.**, Influence of thermal decomposition on the mechanical properties of magnesia-stabilized cubic zirconia, *J. Am. Ceram. Soc.*, 66, 358, 1983.
28. **Rossell, H. J. and Hannink, R. H. J.**, The phase $Mg_2Zr_5O_{12}$ in MgO partially stabilized zirconia, in *Science and Technology of Zirconia II*, Advances in Ceramics, Vol. 12, Claussen, N., Ruhle, M., and Heuer, A. H., Eds., American Ceramic Society, Columbus, Ohio, 1984, 139.

29. **Hannink, R. H. J. and Swain, M. V.**, Magnesia-partially stabilized zirconia: the influence of heat treatment on thermomechanical properties, *J. Aust. Ceram. Soc.*, 18, 53, 1982.

30. **Farmer, S., Heuer, A. H., and Hannink, R. H. J.**, Eutectoid decomposition of MgO-partially stabilized ZrO$_2$, *J. Am. Ceram. Soc.*, 70, 431, 1987.

31. **Dworak, U., Olapinski, H., and Thamerus, G.**, Mechanical strengthening of alumina- and zirconia-ceramics through the introduction of secondary phases, *Sci. Cer.*, 9, 543, 1977.

32. **Swain, M. V., Hughan, R. R., Hannink, R. H. J., and Garvie, R. C.**, Magnesium-partially stabilized zirconia (Mg-PSZ) microstructure and properties, *Uehida Rokakuho*, 2, 117, 1984.

33. **Swain, M. V. and Hannink, R. H. J.**, R-Curve behavior in zirconia ceramics, in *Science and Technology of Zirconia II*, Advances in Ceramics, Vol. 12, Claussen, N., Ruhle, M., and Heuer, A. H., Eds., American Ceramic Society, Columbus, Ohio, 225, 1984.

34. **Swain, M. V.**, Inelastic deformation of Mg-PSZ and its significance for strength-toughness relationships of zirconia toughened ceramics, *Acta Metall.*, 33, 2083, 1985.

35. **Becher, P. F., Swain, M. V., and Ferber, M. K.**, Relation of transformation temperature to the fracture toughness of transformation toughened ceramics, *J. Mater. Sci.*, 22, 76, 1987.

36. **Swain, M. V., Garvie, R. C., Hannink, R. H. J., Hughan, R. R., and Marmach, M.**, Material, development and evaluation of partially stabilized zirconia for extrusion die applications, *Proc. Br. Ceram. Soc.*, 32, 343, 1982.

37. **Sturhahn, H. H., Dawilhl, W., and Thamerus, G.**, Anwendungsmoglichkeiten und Werkstoffeigenschaften von Zirconoxid-Sintererzeugnissen, *Ber. Dtsch. Keram. Ges.*, 52, 59, 1975.

38. **Sturhahn, H. H., Thamerus, G., and Eichas, H. C.**, Novel oxidic materials for wire production, *Wire*, 25, 89, 1975.

39. **Li, L. and Dworak, U.**, Einfluss hoher Temperatur auf die Festigkeitseigenschagten von Teilstabilisierten ZrO$_2$-Wekstoffen, *Ber. Dtsch. Keram. Ges.*, 55, 492, 1978.

40. **Ingel, R. P.**, Structure-Mechanical Property Relationships for Single Crystal Yttrium Oxide Stabilized Zirconium Oxide, Ph.D. thesis, The Catholic University of America, Washington, D.C., 1982.

41. **Ingel, R. P. and Lewis, D.**, Lattice parameters and density for Y$_2$O$_3$ stabilized ZrO$_2$, *J. Am. Ceram. Soc.*, 69(4), 325, 1986.

42. **Yamakawa, T., Ishizawa, N., Uematsu, K., Mizutani, N., and Kato, M.**, Growth of yttria partially and fully stabilized zirconia crystals by xenon arc image floating zone method, *J. Cryst. Growth*, 75, 623, 1986.

43. **Lanteri, V., Heuer, A. H., and Mitchel, T. E.**, Tetragonal phase in the system ZrO$_2$-Y$_2$O$_3$, *Science and Technology of Zirconia II*, Advances in Ceramics, Vol. 12, Claussen, N., Ruhle, M., and Heuer, A. H., Eds., American Ceramic Society, Columbus, Ohio, 1984, 118.

44. **Matsui, M., Soma, T., and Oha, L.**, Effect of microstructure on the strength of Y-TZP components, *Science and Technology of Zirconia II*, Advances in Ceramics, Vol. 12, Claussen, N., Ruhle, M., and Heuer, A. H., Eds., American Ceramic Society, Columbus, Ohio, 1984, 371.

45. **Aleksandrov, V. I., Osiko, V. V., Prokhorov, A. M., and Tatarintsev, V. M.**, *Current Topics in Materials Science*, Vol. 1, Kaldis, E., Ed., N. H., Amsterdam, 1978, 421.

46. **Mizutani, N. and Kato, M.**, Crystal growth and some properties of Sm$_2$O$_3$ and Y$_2$O$_3$ stabilized zirconia single crystals, *Uehida Rokakuho*, 2, 27, 1984.

47. **Michel, D., Mazerolles, L., and Perez Y Jorba, M.**, Fracture of metastable tetragonal zirconia crystals, *J. Mater. Sci.*, 18, 2618, 1983.

48. **Michel, D., Mazerolles, L., and Perez Y Jorba, M.**, Polydomain crystals of single phase tetragonal ZrO$_2$: structure, microstructure and fracture toughness, in *Advances in Ceramics*, Vol. 12, American Ceramic Society, Columbus, Ohio, 1984, 131.

49. **Lefevre, J., Collongues, R., and Perez Y Jorba, M.**, On the cubic-tetragonal transition in systems zirconia — rare earth oxides, *C. R. Acad. Sci.*, 249, 2329, 1959.

50. **Mecartney, M. L., Donlon, W. T., and Heuer, A. H.**, Plastic deformation in CaO-stabilized ZrO$_2$, *J. Mater. Sci.*, 15, 1063, 1980.

51. **Hannink, R. H. J. and Swain, M. V.**, Induced plastic deformation of zirconia, in *Plastic Deformation of Ceramics II*, Tressler, R. E. and Bradt, R. C., Eds., Plenum Press, New York, 1984, 695.

52. **Sakuma, T., Yoshizawa, Y., and Suto, H.**, The microstructure and mechanical properties of yttria stabilized zirconia prepared by arc-melting, *J. Mater. Sci.*, 20, 2399, 1985.

53. **Hannink, R. H. J. and Swain, M. V.**, A mode of deformation in partially stabilized zirconia, *J. Mater. Sci.*, 16, 1428, 1981.

54. **Faber, K. T. and Evans, A. G.**, Crack deflection processes. I. Theory, II. Experiment. *Acta Metall.*, 31, 565, 1983.

55. **Virkar, A. V. and Matsumoto, R. L. K.**, Ferroelastic domain switching as a toughening mechanism in tetragonal zirconia, *J. Am. Ceram. Soc.*, 69, C224, 1986.

56. **Evans, A. G. and Cannon, R. M.**, Overview No. 48: toughening of brittle solids by martensitic transformations, *Acta Metall.*, 34, 761, 1986.

57. **Larsen, D. C. and Adams, J. W.**, Long Term Stability and Properties of Zirconia Ceramics for Heavy Duty Diesel Engine Components, Report DOE/NASA/0305-1 National Aeronautics and Space Administration CR-174943, Lewis Research Center, Cleveland, Sept. 1985, 258.

58. **Gupta, T. K., Bechtold, J. H., Kuznicki, R. C., Cadolf, L. H., and Rossing, B. R.**, Stabilization of the tetragonal phase in crystalline zirconia, *J. Mater. Sci.*, 12, 2421, 1977.

59. **Gupta, T. K.**, Sintering of tetragonal zirconia and its characteristics, *Sci. Sintering*, 10, 205, 1978.

60. **Ruhle, M., Claussen, N., and Heuer, A. H.**, Microstructural studies of Y_2O_3-containing tetragonal ZrO_2 polycrystals (Y-TZP), in *Advances in Ceramics*, Vol. 12, American Ceramic Society, Columbus, Ohio, 352, 1984.

61. **Gross, V. and Swain, M. V.**, Mechanical properties and microstructure of sintered and hot isostatically pressed yttria-partially stabilized zirconia, *J. Aust. Ceram. Soc.*, 22, 1, 1986.

62. **Kobayashi, K., Kuwajima, H., and Misaki, T.**, Phase change and mechanical properties of ZrO_2-Y_2O_3 solid electrolyte after aging, *Solid State Ionics*, 3/4, 489, 1981.

63. **Shigeto, Y., Fuseki, T., and Igarashi, N.**, Transformation Toughened Zirconia (Y-PSZ) Ceramics "TASZIC", Tech Paper No. 83052, Toshiba Ceramics Company, Yokohama, Japan, 1983.

64. **Tsukuma, K., Kubota, Y., and Tsukidate, T.**, Thermal and mechanical properties of Y_2O_3-stabilized tetragonal zirconia polycrystals, in *Advances in Ceramics*, Vol. 12, American Ceramic Society, Columbus, Ohio, 382, 1984.

65. **Swain, M. V.**, Grain size dependence of toughness and transformability of 2 mole % Y-TZP ceramics, *J. Mater. Sci. Lett.*, 5, 1159, 1986.

66. **Sato, T. and Shimada, M.**, Transformation of yttria doped tetragonal ZrO_2 polycrystals by annealing in water, *J. Am. Ceram. Soc.*, 68, 356, 1985.

67. **Michalske, T. A. and Freiman, S. W.**, A molecular mechanism for stress corrosion in vitreous silica, *J. Am. Ceram. Soc.*, 66, 23, 1983.

68. **Schmauder, S. and Schubert, H.**, Significance of internal stresses for the martensitic transformation in yttria stabilized tetragonal zirconia polycrystals during degradation, *J. Am. Ceram. Soc.*, 69, 543, 1986.

69. **Lange, F. F., Marshall, D. B., and Porter, J. R.**, *Advances in Structural Ceramics*, Becher, P. F., Swain, M. V., and Somiya, S., Eds., MRS, Pittsburgh, 1987.

70. **Masaki, T.**, Mechanical Properties of Y-PSZ after aging at low temperature, *Int. J. High Tech. Ceramics*, 2, 85, 1986.

71. **Nakanishi, N., Shigematsu, T., Sugimura, T., and Okinaka, H.**, Mechanism of isothermal propagation of martensite and generation of microcracks in Y_2O_3 partially stabilized zirconia ceramics, *Advances in Ceramics*, Vol. 24, American Ceramic Society, Columbus, Ohio, in press.

72. **Claussen, N.**, Microstructural design of zirconia toughened ceramics (ZTC), in *Advances in Ceramics*, Vol. 12, American Ceramic Society, Columbus, Ohio, 1984, 325.

73. **Tsukuma, K. and Shimada, M.**, Strength, fracture toughness and Vickers hardness of CeO_2-stabilized tetragonal ZrO_2 polycrystals (Ce-TZP), *J. Mater. Sci.*, 20, 1178, 1985.

74. **Tsukuma, K.**, Mechanical properties and thermal stability of CeO_2 containing tetragonal zirconia polycrystals, *Am. Ceram. Soc. Bull.*, 65, 1386, 1986.

75. **Lange, F. F.**, Transformation toughening, *J. Mater. Sci.*, 17, 255, 1982.

76. **Coyle, T. W., Coblenz, W. S., and Bender, B. A.**, Toughness, strength and microstructures of sintered CeO_2-doped ZrO_2 alloys (Abstr.), *Am. Ceram. Soc. Bull.*, 62, 966, 1983.

77. **Coyle, T. W.**, Ph.D. thesis, Massachusetts Institute of Technology, Cambridge, Mass., 1985.

78. **Cook, R. F. and Lawn, B. R.**, A modified indentation toughness technique, *J. Am. Ceram. Soc.*, 66, C200, 1983.

79. **Hannink, R. H. J., Muddle, B. C., and Swain, M. V.**, Transformation induced plasticity in tetragonal zirconia polycrystals, *Proc. Austceram 86*, Australian Ceramic Society, Melbourne, 1986, 145.

80. **Swain, M. V., Hannink, R. H. J., and Drennan, J. D.**, Interfacial related properties of partially stabilized zirconia ceramics, *Ceramics Microstructure '86: Interfaces*, in press.

81. **Hannink, R. H. J. and Muddle, B. C.**, Stress induced phase transformations in ceria-zirconia, in *Analytical Electron Microscopy*, Joy, D. C., Ed., San Francisco Press, 1987, 30.

82. **Hannink, R. H. J. and Swain, M. V.**, Metastability of the martensitic transformation in a 12 mol % CeO_2-ZO_2 alloy. I. Deformation and fracture observations, *J. Am. Ceram. Soc.*, in press; **Swain, M. V. and Hannink, R. H. J.**, II. Grinding Studies, *J. Am. Ceram. Soc.*, submitted.

83. Commonwealth Scientific and Industrial Research Organization, Unpublished data.

Chapter 5

MICROSTRUCTURE-MECHANICAL BEHAVIOR OF ZIRCONIA-DISPERSED CERAMICS (ZDC)

I. INTRODUCTION

The basis for discussion in this chapter are some of the materials in the second row of the matrix of zirconia-toughened ceramics (ZTC) as classified by Claussen[1] (see Chapter 1, Figure 6). This row is shown in Figure 1 and consists of six specific possibilities; of these, the major emphasis will be upon the intercrystalline-based and *in situ* reacted materials. The reason for this limitation is that the major research efforts in this field have concentrated in these areas primarily because of simplicity in fabrication, that is, traditional milling-mixing techniques followed by conventional sintering. There is currently a trend to develop more complex fabrication techniques suitable for mass production of highly reactive, premixed powders such as sol-gel, chemical vapor reaction, amorphous rapidly quenched powders, etc. Some of the products of these approaches will be mentioned in this chapter.

To date the largest data set pertains to the zirconia dispersed in oxide systems. The relative inertness of ZrO_2 enables it to be sintered at modest temperatures ($\leq 1600°C$) with a number of other oxide and nonoxide materials with only limited, if any, reactions. The role of various cations as stabilizers of the cubic-fluorite phase of ZrO_2 has been discussed in Chapter 2. A typical example of a MeO-ZrO_2 phase diagram[2] is shown in Figure 2 for Al_2O_3-ZrO_2. This simple binary system indicates the virtual complete compatibility of Al_2O_3 and ZrO_2 over the entire composition range. A eutectic at 1880°C occurs at a composition of approximately 40 wt %. Limited evidence for the solubility of Al_2O_3 in ZrO_2 is known, although Bannister[3] has found this increases in ZrO_2 (Y_2O_3) compositions. More detailed phase diagrams in ZrO_2-binary and ternary systems are available in compendia of such diagrams.[2] Less well defined and available are phase diagrams in the ZrO_2-nonoxide fields. The phase fields are further complicated by the mobility of the oxygen ions in the ZrO_2 and associated changes in stoichiometry. As mentioned in Chapter 2, many zirconia-dispersed ceramic (ZDC) materials are not equilibrated under the processing/sintering conditions employed.

The development of toughened alumina-zirconia composite systems occurred almost simultaneously with the conceptual breakthrough in PSZ systems. Often with the oxide-dispersed system, the addition of zirconia has more than one influence; for instance, it may act as a grain boundary-pinning phase and thereby limit grain growth, and as a grain boundary glassy-phase scavenger. Moreover, the mechanism by which the zirconia acts to toughen the composite is variable and may include transformation toughening, microcrack, and crack deflection mechanisms (see Chapter 3). Indeed it is possible to fabricate a single material with all three mechanisms acting.

The successful development of zirconia-dispersed nonoxide materials has been much slower than for oxide materials. The generally higher sintering temperatures and the degradation associated with the presence of significant oxygen-generating materials, of which zirconia is an excellent example at modest temperatures, have made fabrication more difficult. There have been a number of reports of the successful fabrication of such nonoxide systems, but to date these are very much less evident at the pilot plant/preliminary commercialization stage than the oxide systems. One significant problem has been the reaction of the nonoxide anionic component with zirconia leading to the stabilization or formation of a new nontransformable phase. Another feature of nonoxide systems, particularly silicon based, such as SiC and Si_3N_4, is the large difference in thermal expansion coefficient between the ZrO_2 and matrix. However, as with oxide materials, it has been found that the addition of ZrO_2

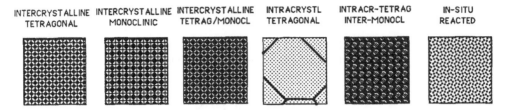

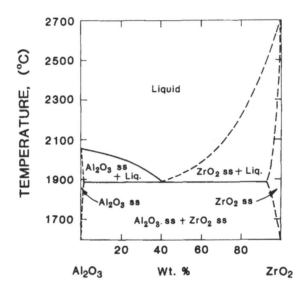

FIGURE 1.   A schematic summary of the simpler zirconia-dispersed ceramic (ZDC) composites. (After Claussen, N., *Science and Technology of Zirconia, II, Advances in Ceramics,* Vol. 12, Claussen N., Ruhle, M., and Heuer, A. H., Eds., American Ceramic Society, Columbus, Ohio, 1984, 325.)

FIGURE 2.   The Al_2O_3-ZrO_2 phase diagram as an illustration of a system in which ZDC materials may be readily fabricated. (After Levin, E. M., Robbins, C. R., and McMurdie, H. F., Phase Diagrams for Ceramics, American Ceramic Society, Columbus, Ohio, 1969.)

often imparts unanticipated benefits such as the removal of the glassy grain-boundary phase and improvement in the high-temperature strength and creep resistance.

II. ALUMINA-ZIRCONIA (ZIRCONIA-TOUGHENED ALUMINA — ZTA)

The incorporation of ZrO_2 into oxide ceramics to deliberately toughen them was undertaken shortly after the advances in PSZ work. Claussen et al.[4,5] showed that the inclusion of unstabilized zirconia could lead to retention of t-ZrO_2 in the fabricated body, provided the particle size was small enough. Improved toughness of Al_2O_3-ZrO_2 systems was possible with both t- and m-ZrO_2 incorporated in the composite body. The mechanisms responsible for the toughening increments are still in dispute, but both transformation and microcrack toughening are considered the most likely mechanisms.[6,7] The amount of unstabilized ZrO_2 that can be added is typically <20 vol %. More recent work in Al_2O_3-ZrO_2 has centered on increasing the ZrO_2 content by partial stabilization of the zirconia with Y_2O_3, CeO_2, and more recently TiO_2, as this allows a complex range of composites to be fabricated containing any volume percent of ZrO_2. The other development in this area has been to combine a

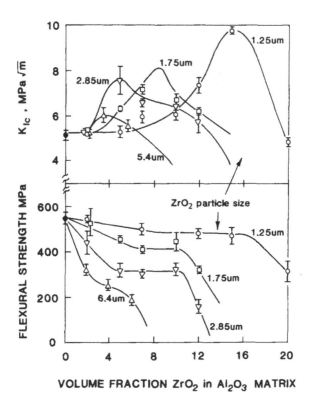

FIGURE 3. Fracture toughness, K_{IC}, and flexural strength dependence of Al_2O_3-ZrO_2 composites on volume fraction and particle size of m-ZrO_2. (After Claussen, N., Steeb, J., and Pabst, R. F., *Am. Ceram. Soc. Bull.*, 59, 49, 1976.)

third phase such as TiC or SiC (in the powder or whisker form) to improve or modify the hardness-elastic modulus, thermal-electrical conductivity or the toughness wear resistance of such bodies. In this section, developments and properties of Al_2O_3-ZrO_2 composites in the following three areas (unstabilized, partially stabilized, and mixed zirconia nonoxide systems) will be addressed. The area of whisker reinforced systems is currently under intense investigation, but will receive only a cursory treatment.

A. Al_2O_3-Unstabilized ZrO_2

Claussen et al.[4,5] fabricated a range of Al_2O_3-ZrO_2 composites with both varying ZrO_2 content, up to 20 vol % and particle sizes ranging from 1.25 to 6.4 μm prior to fabrication. The bodies were hot pressed in graphite dies under vacuum at a pressure of 40 MPa for 30 min at 1500°C. Strengths were measured in three-point bending on a 28 mm span and fracture toughness with the single-edge notched-bend (SENB) technique. The actual particle sizes of the zirconia in the fabricated bodies are somewhat larger than the admixed size. The results are shown in Figure 3.

There is a distinct difference in the volume-fraction dependence of results for K_{IC} and strength. The K_{IC} results all pass through a maximum with increasing volume fraction of ZrO_2. The maxima occur at higher volume fractions with decreasing particle size. The strength results show only a deterioration with volume fraction of zirconia.

The degradation in strength is most rapid with the largest ZrO_2 particles; for the material containing 1.25 μm particles, the strength begins to decrease at volume fractions above 16 vol % the peak in toughness. Claussen et al.[5] provided transmission electron microscope

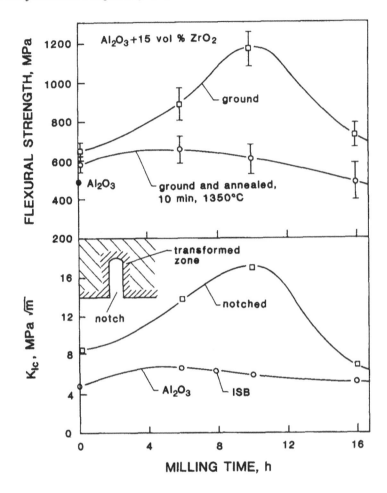

FIGURE 4.   Flexural strength and fracture toughness of a Al_2O_3-15 vol % ZrO_2 composite with milling time (particle size) of unstabilized ZrO_2. (After Claussen, N. and Jahn, J., *J. Am. Ceram. Soc.*, 63, 228, 1980.)

(TEM) evidence for the formation of microcracks about individual m-ZrO_2 particles and both microcracking and crack branching about an extended crack in such materials. They also noted that the macroscopic fracture surfaces were much rougher in the ZrO_2-containing material than in pure Al_2O_3 or Al_2O_3 containing c-$ZrO_2(Y_2O_3)$. Another feature these authors observed was the ZrO_2 particle volume fraction dependence of the M_s and M_f and to much lesser extent A_s and A_f temperatures. A number of authors[8-10] have proposed theoretical models based on microcracking to explain the observations in Figure 3. This toughening mechanism has been discussed in detail in Chapter 3, Section IV.B.

Subsequent studies by Claussen[11] indicated that t-ZrO_2 could be retained in the Al_2O_3 matrix by finely milling (attritor milling) the zirconia powder. The critical particle size of unstabilized zirconia particles is a function of volume fraction and has been discussed in Chapter 2. The effect of milling time was to reduce the ZrO_2 particle size and after 8 hr milling for the 15 vol % ZrO_2 material only 20 % (by volume) of the ZrO_2 remained as monoclinic. Increasing the volume fraction of ZrO_2 to 20%, Claussen and Jahn[12] showed that even after 18 hr of attritor milling the majority of the material remained m-ZrO_2. The influence of milling time on the strength of ground bars before and after annealing is shown in Figure 4 for an Al_2O_3-15 vol % ZrO_2 material. Also shown in this figure are SENB values of K_{IC} and more realistic values of the K_{IC} determined by the retained strength after indentation

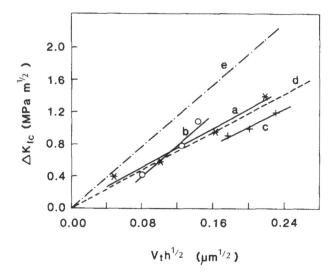

FIGURE 5. Fracture toughness increment ΔK_{IC} of various Al_2O_3-ZrO_2 composites on transformation zone size (h) and volume fraction of transformed t-ZrO_2 (V_t). The latter two quantities were determined from X-ray analysis of the fracture surface. Most of the data was obtained from composites containing essentially t-ZrO_2, some of which were partially stabilized with Y_2O_3. Compositions of the various materials are (a) Al_2O_3 + x^v/$_o$ (ZrO_2 + 2 ^{m}/$_o$ Y_2O_3), (b) Al_2O_3 + 7.5 ^{v}/$_o$ ZrO_2, (c) Al_2O_3 + 15 ^{v}/$_o$ ZrO_2, and (d) and (e) theoretical predictions. (After Kosmac, T., Swain, M. V., and Claussen, N., *Mater Sci. Eng.*, 71, 59, 1985.)

techniques (ISB) for the same materials.[13] Details concerning the actual particle sizes have been given by Heuer et al.[14]

The large difference in strength between ground and ground-plus annealing at 1350°C for 10 min was related to the surface compressive stresses developed during grinding and relieved upon annealing. The maximum difference between these two sets of results occurs when the majority of the t-ZrO_2 particles are readily transformed by grinding. This occurs after approximately 10 hr of milling and corresponds with the condition for maximum transformation zone size.[15]

It has been found that the depth of the transformation zone produced by grinding depends on the volume fraction of ZrO_2, the grinding pressure, and the zirconia particle size.[16,17] The strength is a function of these parameters and the influence and magnitude of these surface stresses are discussed in Chapter 6. The results of ISB K_{IC} measurements show that the maximum K_{IC} occurs after only 6 hr of milling when there remains approximately 40% m-ZrO_2. If only transformation toughening were contributing to the K_{IC}, then the peak value should occur after 10 hr of milling or at the greatest volume fraction of transformable t-ZrO_2 present in the materials (see Chapter 3). These observations and further studies by Kosmac et al.[6] indicate that both microcracking and transformation toughening occur and appear to be additive.

According to theoretical predictions in Chapter 3, the toughness should scale with the volume fraction of transformable t-ZrO_2 and the square root of the depth of the transformed zone (h) surrounding the crack. Observations in support of this premise are shown in Figure 5 for a range of Al_2O_3-ZrO_2 composites containing essentially t-ZrO_2.[6] Stabilization of the t-ZrO_2 phase was achieved in higher volume fractions of ZrO_2 by addition of 2 mol % Y_2O_3. The theoretical estimate of the toughening increment has been determined from expressions given in Chapter 3 with $\Delta v/v$ (or e^T) = 0.026. This value is much less than the unconstrained

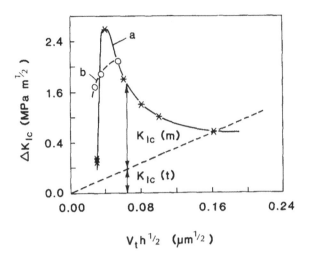

FIGURE 6. Fracture toughness increment ΔK_{IC} of Al_2O_3-ZrO_2 composites vs. transformed zone size for materials containing t- and m-ZrO_2. The total increment appears to be a summation of transformation and microcrack toughening. Compositions of materials are as follows (a) Al_2O_3 + x (ZrO_2 + 2 $^m/_o$ Y_2O_3), (b) Al_2O_3 + 15 $^v/_o$ (ZrO_2 + x Y_2O_3), (c) Al_2O_3 + 15 $^v/_o$ (ZrO_2 + x HfO_2) (d) Al_2O_3 + 15 $^v/_o$ ZrO_2, and (e) Al_2O_3 + 7.5 $^v/_o$ ZrO_2. (After Kosmac, T., Swain, M. V., and Claussen, N., *Mater. Sci. Eng.*, 71, 59, 1985.)

volume dilation on transformation and incorporates thermal expansion mismatch between the Al_2O_3 matrix and the t-ZrO_2.

The relationship between K_{IC} and transformed zone size h becomes more complex for the composite materials containing substantial quantities of m-ZrO_2. The discrepancy between transformation-toughening theory and experimental results[6] for three composites is shown in Figure 6. The toughening increment ΔK_{IC} is again plotted against the volume fraction of t-ZrO_2 and transformation zone size. The observed toughness values are much higher than anticipated on the basis of transformation toughening indicating that another mechanism is operative in these composites. The maximum K_{IC} does not correspond to the highest transformation component, but it appears that microcrack toughening also occurs. Kosmac et al.[6] suggest, as shown in Figure 6, that the total toughness increment is the sum of transformation and microcrack toughening.

TEM evidence for microcrack toughening in Al_2O_3-ZrO_2 composites has been presented by Ruhle et al.[7] These authors, in a study of composites containing 15 vol % ZrO_2 prepared by sintering and subsequent hot isostatic pressing (HIP) confirmed the observations of Kosmac et al.[6] in Figures 5 and 6. They found that in materials containing t-ZrO_2, a transformed zone existed up to 8 μm from the crack tip, with 50% transformed at ~5 μm. The larger t-ZrO_2 particles were more likely to have transformed and particles <0.2 μm were not observed to transform. In materials containing mainly m-ZrO_2 particles, it was found that microcracking was observed at 6 μm, but not at 20 μm from the crack tip. The density of microcracks increased with proximity to the crack and was more likely to surround larger than smaller m-ZrO_2 particles.

Other studies of the variation in K_{IC} of Al_2O_3-ZrO_2 composites with zirconia content by Lange[18] and Becher and Tennery[19] show that the peak toughness[20] in the materials occurs at approximately 10 vol % ZrO_2.

At higher volume fractions of zirconia, it was found difficult to retain the t-ZrO_2 phase

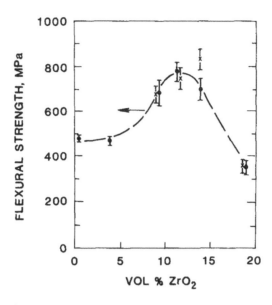

FIGURE 7. Dependence of the flexural strength of Al_2O_3-ZrO_2 composites on volume fraction of unstabilized ZrO_2. The results show only a minor difference between ground (x) and annealed (·) strength. (After Becher, P. F., *J. Am. Ceram. Soc.*, 64, 37, 1981.)

in the sintered body. The peak strength of these materials (Figure 7) occurs at slightly higher ZrO_2 contents than for peak toughness.[20] The lower toughness values reported by Becher[20] are comparable to those shown in Figure 3 and indicate care must be exercised when using the SENB techniques to determine toughness. The SENB technique tends to overestimate K_{IC} in fine-grained materials because the notch radius is usually larger than the grain size. This is further exaggerated by the compressive zone at the notch introduced into transformation-toughened ceramics.[13]

Other techniques for fabricating Al_2O_3-ZrO_2 composites, apart from milling, include the sol-gel method,[20] hydrothermal oxidation, melting and rapid quenching,[22] and CVD.[23,24] The latter studies[23,24] have shown some interesting effects of HIP on the flexural strength. The CVD technique enables high-purity, very homogeneous materials to be fabricated and the complete retention of the t-ZrO_2 up to ~15 vol % additions of ZrO_2. This feature is shown in Figure 8 which compares as-sintered and fractured surfaces of sintered and HIP composites. The difference in t-ZrO_2 content between the sintered and fractured surfaces provides an indication of the depth of the transformed zone about the crack. Unlike the results of Claussen et al.,[5] Lange,[18] and Becher and Tennery,[19] Hori et al.[23,24] find that the K_{IC} of their materials determined using the ISB and indentation-fracture technique increases monotonically with ZrO_2 content and is independent of its phase. These observations are similar to the most recently reported by Ruhle et al.[7] However, HIP does significantly improve the strength of the sintered body,[24] but only for materials containing t-ZrO_2 as shown in Figure 9. The influence of HIP on the K_{IC} is marginal. These observations suggest that the increase in strength is related to a decrease in flaw size for the HIP t-ZrO_2 containing material. The bodies containing m-ZrO_2 are not strengthened because of microcrack development during t → m-ZrO_2 transformation upon cooling.

Becher[20] has shown that the addition of unstabilized ZrO_2 to Al_2O_3 greatly improves its thermal-shock resistance. Plots of retained strength vs. temperature difference after plunging into boiling water[20] are shown in Figure 10. The critical ΔT_c for the peak-strength material,

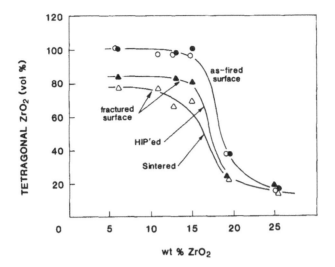

FIGURE 8.   Dependence of the t-ZrO$_2$ content of Al$_2$O$_3$-ZrO$_2$ composites on the as-fired and fracture surface of sintered and HIP materials as a function of weight % of ZrO$_2$. (After Hori, S., Yoshimura, M., and Somiya, S., *J. Am. Ceram. Soc.*, 69, 169, 1986.)

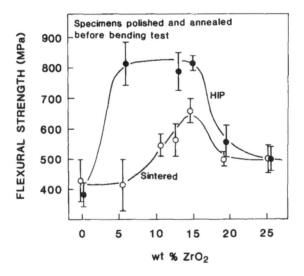

FIGURE 9.   Flexural strength of sintered and HIP- Al$_2$O$_3$-ZrO$_2$ composites as a function of weight % ZrO$_2$ of CVD derived powders. (After Hori, S., Yoshimura, M., and Somiya, S., *J. Am. Ceram. Soc.*, 69, 169, 1986.)

which contained 11.5 $^v/_o$ ZrO$_2$, had a threefold increase over that of the matrix material. The critical ΔT_c appears to scale reasonably well with the peak strength as anticipated from figure of merit thermal-stress resistance predictions.[25] Claussen and Ruhle[26] also reported a significant improvement of the thermal-shock resistance of Al$_2$O$_3$-ZrO$_2$ composites over that of Al$_2$O$_3$, but with a strong dependence of ΔT_c upon water bath temperature. An additional strategy for improving the thermal-shock resistance of ATZ materials was proposed by Claussen and Jahn[12] of using the addition of different size populations of ZrO$_2$ to decrease the effective thermal expansion coefficient.

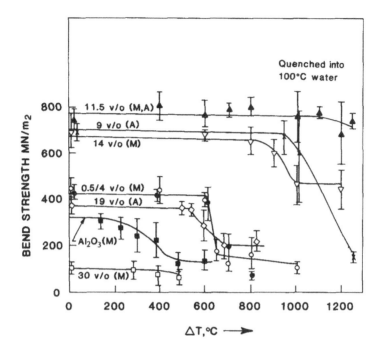

FIGURE 10.   Retained strength of various Al₂O₃-ZrO₂ composites after thermal shocking in boiling water. The results show minimal difference between machined (M) and annealed (A) specimens. (After Becher, P. F., *J. Am. Ceram. Soc.*, 64, 37, 1981.)

The microstructures and ZrO_2 phase composition of Al_2O_3-ZrO_2 composites are strongly dependent upon processing conditions. The critical size of ZrO_2 particles for transformation to m-ZrO_2 on cooling has been the subject of a number of studies[14,26,27] and some controversy exists as to the mechanism responsible for this transformation. The observations of Claussen and Ruhle,[26] for the variation in the M_s temperature with ZrO_2 intergranular particle size, are shown in Figure 11 for Al_2O_3-16 vol % ZrO_2 composites. More recent studies by Green[28] show that the critical ZrO_2 particle size for transformation is dependent upon the volume fraction of ZrO_2. These observations have been confirmed by Garvie[27] who finds minimal transformation takes place below 400 to 500°C. Green[28] observed that the critical ZrO_2 size changes from ≥ 2 μm at 5 vol % ZrO_2 to ~0.7 μm at 20 vol %. Associated with a higher volume fraction of m-ZrO_2, microcracking and a significant reduction in Young's modulus are observed[28] (Figure 12).

Apart from increasing the toughness of Al_2O_3, the addition of ZrO_2 imparts benefits in terms of microstructure control. Green[28] noted that in excess of 5 vol % ZrO_2 prevented abnormal grain growth of the Al_2O_3 when heat treated at 1650°C. A more definitive study by Lange and Hirlinger,[29] with the incorporation of finer ZrO_2, confirmed the observations by Green[28] and found grain growth to be controlled with as little as 5 vol % ZrO_2. Less than 5 vol % addition would not control occasional abnormal grain growth. Lange and Hirlinger[29] proposed that the ZrO_2 inclusions moved with the Al_2O_3 four grain junctions during grain growth and that growth of the ZrO_2 inclusions occurred by coalescence. The inclusions exerted a dragging force at the four grain junctions to limit grain growth and the resulting grain size was inversely proportional to the volume fraction of the inclusions.

Kibbel and Heuer[30,31] have examined the factors controlling the ripening of inter- and intragranular ZrO_2 particles in alumina-zirconia composites. They find that the growth rate of intergranular species is controlled by Ostwald ripening via grain-boundary diffusion

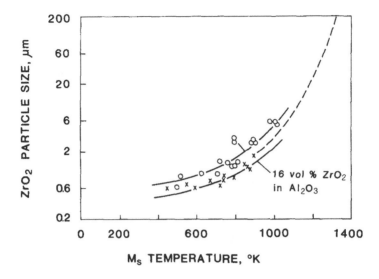

FIGURE 11.   Variations in the M_s temperature of Al_2O_3-ZrO_2 composites as a function of ZrO_2 particle size. (After Claussen, N. and Ruhle, M., in *Science and Technology of Zirconia, Advances in Ceramics*, Vol. 3, Heuer, A. H. and Hobbs, L. W., Eds., American Ceramic Society, Columbus, Ohio, 1981, 137.)

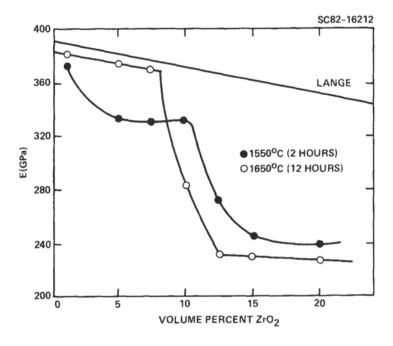

FIGURE 12.   Dependence of Young's modulus of Al_2O_3 unstabilized ZrO_2 composites upon volume fraction of ZrO_2. The sharp reduction at 10 vol % ZrO_2 marks the onset of microcracking in the specimen. (After Green, D. J., *J. Am. Ceram. Soc.*, 65, 610, 1982.)

(particle size proportional to time$^{1/4}$). Intragranular particles, usually observed in sol-gel derived materials, did not conform to a classic Ostwald ripening. The rate of growth appears to be dependent upon location with those particles adjacent to grain boundaries growing most rapidly, leading to denuded zones adjacent to Al_2O_3 grain boundaries.

The major reported application of Al_2O_3-ZrO_2 composites has been metal cutting tools.

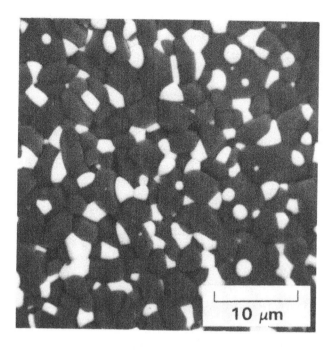

FIGURE 13. Typical microstructure of an Al$_2$O$_3$-ZrO$_2$ when viewed in the scanning electron microscope (SEM) after polishing and thermal etching. The lighter phase is ZrO$_2$.

Alumina-cutting tools have been available for at least 30 years; however, they were suitable only for continuous shallow high-speed cutting of various metals because of their fragility. As recently discussed by Garvie,[32] the influence of the addition of zirconia and the improved toughness has improved the impact resistance and thereby the range of cutting applications. A typical cutting tool contains 16 vol % ZrO$_2$, with approximately equal amounts of m- and t-ZrO$_2$. An SEM micrograph of a polished and thermally etched sample is shown in Figure 13. The lighter phase present at the grain boundary triple points is zirconia. A comparison between the performance of Al$_2$O$_3$-ZrO$_2$ composites, with or without Y$_2$O$_3$ to partially stabilize the t-ZrO$_2$ phase, is available in an article by Rieter.[33]

B. Al$_2$O$_3$-TZP Materials

The addition of Y$_2$O$_3$ to ZrO$_2$, to partially stabilize the t-ZrO$_2$ phase and to enable less-restricted fabrication conditions, has been studied by a number of authors.[34-37] As discussed by Lange,[35] and in more detail in Chapter 2, the role of Y$_2$O$_3$, CeO$_2$, and other ReO is to reduce the chemical driving force for the t $\rightarrow$ m transformation and, thereby, enable larger ZrO$_2$ particles to be retained in the t-ZrO$_2$ form. In this manner, it is possible to increase the t-ZrO$_2$ volume fraction well beyond the 15 to 20 $^v/_o$ possible in the unstablized ZrO$_2$-Al$_2$O$_3$ systems. Lange[35] demonstrated that materials across the complete ZrO$_2$-Al$_2$O$_3$ range could be fabricated with only t-ZrO$_2$ symmetry by the addition of 2 $^m/_o$ (mol %) Y$_2$O$_3$ to the ZrO$_2$. The critical size of the t-ZrO$_2$ grains was somewhat larger in the mixed composite than the Y-TZP material because of the higher elastic constraint of the composite (E Al$_2$O$_3$ > E ZrO$_2$). This is somewhat offset by the mismatch of the thermal expansion coefficient of the ZrO$_2$ and Al$_2$O$_3$. Lange[35] found that the elastic modulus and hardness of the mixed composite fabricated by hot pressing at 1600°C appeared to follow a rule of mixture prediction independent of whether the zirconia was t or c phase[35] (Figure 14). The fracture toughness, however, was very strongly dependent upon the form of zirconia and showed a peak in

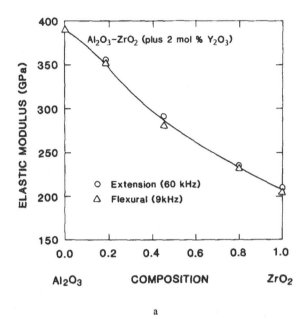

a

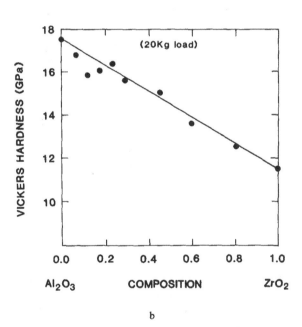

b

FIGURE 14. (a) Young's Modulus vs. composition for Al_2O_3-ZrO_2 (2 mol % Y_2O_3) composites. (b) Variation in Vickers hardness, measured with a 20 kg load, for hot pressed Al_2O_3-ZrO_2 (2 mol % Y_2O_3) composites. (After Lange, F. F., *J. Mater. Sci.*, 17, 255, 1982.)

toughness for the t-ZrO_2-containing material at approximately equal volume fractions of the end members[35] (Figure 15).

These observations have been confirmed by Becher and Tennery[19] at the alumina-rich end of the composite. Both studies[19,35] found that the toughness of t-ZrO_2-containing composites was strongly dependent upon temperature. As discussed in Chapter 3, Section III.E, the magnitude of transformation toughening decreases with increasing temperature. The

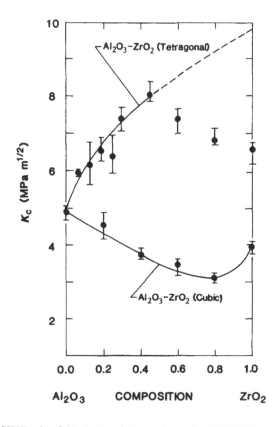

FIGURE 15. Critical stress intensity factor for Al₂O₃-ZrO₂ composite with t-ZrO₂ (2 mol % Y₂O₃) and c-ZrO₂ (7.5 mol % Y₂O₃). (After Lange, F. F., *J. Mater. Sci.*, 17, 255, 1982.)

temperature dependence of K_{IC} for a range of Al₂O₃-ZrO₂ (Y₂O₃) composites from −200°C to 600°C is shown in Figure 16. The linear dependence of K_{IC} with temperature occurs because the chemical driving force for transformation, ΔF_{CH} is directly dependent upon temperature (see Chapter 3).

Materials in the zirconia-rich region of the phase field have been studied in some detail by Tsukuma et al.[36,37] They have shown that the addition of up to 40 wt % of Al₂O₃ to 2 or 3 mol % Y-TZP leads to extremely high strengths with values as high as 2.4 GPa being measured after HIP. Examples of the influence of alumina to the strength of 2 and 3 mol % Y-TZP are shown in Figure 17.

The strength of the composite material increases substantially with the addition of Al₂O₃ with a maxima at 20 wt % Al₂O₃. The strength maxima is only weakly dependent upon Y₂O₃ content in the range of 2 to 4 mol % (Figure 18). The strengths shown in Figure 18 were determined from 3-point bend tests on spans of 4 × 3 mm on a 30 mm span. The strength of these materials is very dependent on processing conditions. Lange et al.,[38] in a series of papers, have examined the critical role of processing in the fine-grained materials if high strengths are to be achieved. A number of strength-limiting flaws have been identified from SEM examination of fracture origins including agglomerates, inhomogenous powders containing the odd large-grain, poor mixing, organic fibers, etc. Some of these have been shown and discussed in Chapter 1, Figure 7. Strategies for the the elimination of these various potential failure sites have also been identified by Lange et al.[38]

The fracture toughness of these materials shows almost the inverse behavior to the strength with addition of alumina (Figure 19). The increased strength of the composite systems, with

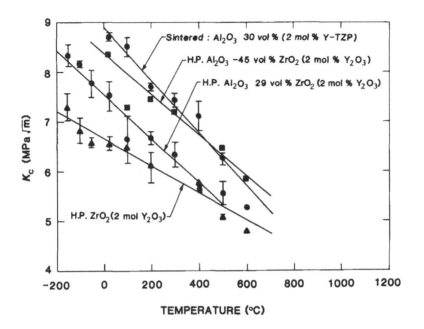

FIGURE 16.   Temperature dependence of the critical stress intensity factor for a number of Al_2O_3-ZrO_2 composites containing varying quantities of ZrO_2 with 2 mol % Y_2O_3. Results obtained from the Vicker's indentation technique. (After Lange, F. F., *J. Mater, Sci.*, 17, 255, 1982.)

up to 40 wt % Al_2O_3, appears to be related to a decrease in flaw size, as is evident in SEM observations of fracture origins as shown in Figure 20 for as-sintered and sintered-plus-HIP material. Another factor leading to the increased strength is that the critical stress to initiate the t → m transformation increases with decreasing toughness (see Chapter 3, Figure 22b).

The influences of temperature on the strength of materials containing 3 mol % Y_2O_3 and with up to 40 wt % Al_2O_3 is shown in Figure 21. Similar results were obtained with 2 $^m/_o$ Y_2O_3-containing materials; the strength at 1000°C was just below 1 GPa. The rapid reduction in strength with temperature closely follows the results of Lange[35] in that the strength reaches a plateau at 400°C (ΔF_{CH} → 0) with the exception of the pure 2 mol % TZP material which continues to decline with increasing temperature. Other beneficial effects of the addition of Al_2O_3 are the increased thermal-shock resistance, increased stability of the t-ZrO_2 in moist environments, and increased creep resistance at 1200°C. The critical thermal-shock resistance ΔT_c for bars plunged into water increases from ~200°C for single phase Y-TZP to 480°C for material containing 20 and 40 wt % Al_2O_3. The retained strength of bars for thermal-shock conditions with $\Delta T \geqslant \Delta T_c$ are in some instances zero (Figure 22). The bending strength of bars after aging at 300°C for the pure 2 mol % Y-TZP and materials containing 20 wt % Al_2O_3 after normal sintering (NS) and HIP is shown in Figure 23. The rate of development of m-ZrO_2 on the surface is much slower for the HIP 20 wt % Al_2O_3 material and the depth of the transformation zone is also much less.

The creep resistance at 1200°C is also considerably improved by the addition of Al_2O_3. The mechanism by which the alumina increases the properties is still being sought, although limited TEM observations suggest that it does influence the wettability of silica-rich grain boundary on the ZrO_2-ZrO_2 vs. ZrO_2-Al_2O_3 interface. Other parameters that are modified by the addition of alumina include a reduction in thermal expansion coefficient, an increase in the thermal diffusivity, and a decrease in density. These parameters, combined with the ability to modify hardness and Young's modulus, provide synergisms for the design of transformation-toughened materials for specific applications.

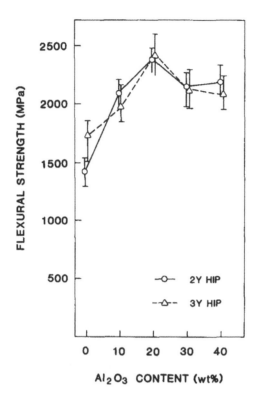

FIGURE 17. Dependence of flexural strength of 2 and 3 mol % Y_2O_3-ZrO_2 materials with alumina addition. All specimens were HIP after sintering. Specimen dimensions $40 \times 3 \times 4$ mm tested in 3-point bending. (After Tsukuma, K., Ueda, K., Matsushita, K., and Shimada, M., *J. Am. Ceram. Soc.*, 68, C-56, 1985; Tsukuma, K,. Ueda, K., Shimada, M., *J. Am. Ceram. Soc.*, 68, C-4, 1985.)

Lange[35] has also explored the effect of stabilization of ZrO_2 with CeO_2 over the range 11 to 22 mol % on the K_{IC} of Al_2O_3-30 vol % ZrO_2 and finds a peak at approximately 12 mol % (8 MPa $\sqrt{m}$), then a gradual decrease with an increase in CeO_2 content. The decrease is anticipated because of the decrease in ΔF_{CH} with an increase in CeO_2 content. Other factors that will influence the K_{IC} are the grain size of the ZrO_2, as well as the stabilizer content. As discussed in Chapter 3, increasing t-ZrO_2 grain size or decreasing stabilizer content will increase the magnitude of the transformation-toughening increment.

III. β″-ALUMINA-ZIRCONIA

Sodium β″-alumina has attracted much interest recently because of its application as a solid electrolyte membrane in the sodium-sulfur battery and the sodium heat engine.[39,40] The sodium-sulfur battery operates at a temperature of about 350°C. Apart from the general fragility of β″-alumina, which has limited potential application as a power source in transportation systems, it has also been recognized that it undergoes strength degradation in service. Two failure modes have been identified by De Jonghe et al.,[41] the more serious of which appears to be dendrite formation during charging. This type of degradation is associated with the formation of liquid-sodium-filled cracks. These cracks, which appear to initiate at the Na$^+$ exit surface, propagate through the thickness of the membrane, leading to electrical breakdown due to short circuit.

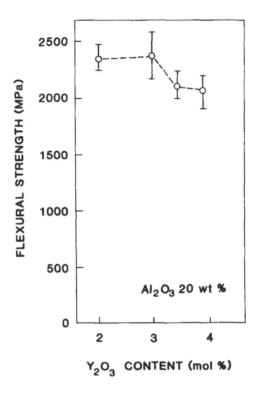

FIGURE 18.   Strength dependence of ZrO_2-20 wt % Al_2O_3 composites on Y_2O_3 content of the zirconia. All specimens HIP. (After Tsukuma, K., Ueda, K., Matsushita, K., and Shimada, M., *J. Am. Ceram. Soc.*, 68, C-56, 1985; Tsukuma, K., Veda, K., Shimada, M., *J. Am. Ceram. Soc.*, 68, C-4, 1985.)

A number of models have been proposed to explain the degradation of the electrolyte,[42-44] a common feature to all of them is the concept of sodium under pressure in the cracks. One significant theoretical result has been the development of an expression for the critical current density i_{cr} during the charging mode, below which sodium penetration will not occur. Virkar[45] has predicted that the current density above which degradation occurs is proportional to the fourth power of K_{IC} (i.e., $i_{cr} \propto K_{IC}^4$). Thus, modest improvements in K_{IC} should lead to dramatic increases in i_{cr}. This would have significant influence on the economic advantages of the sodium-sulfur battery system. For instance, not only would the battery life be enhanced, but the membrane thickness could be reduced and the charging time drastically reduced. All three factors are expected to be important for the possible viability of electric-powered transportation systems.

A number of papers have reported on the practicability of strengthening and toughening of β″-alumina through the incorporation of ZrO_2.[46-50] The approaches adopted by these various authors, in the manner of incorporating zirconia into the β″-alumina, are significantly different and are probably reflected in the properties reported. Generally a small amount of Li_2O (0.75 wt %) was added to the β″-alumina to stabilize this phase. Viswanathan et al.[46] wet milled in acetone β″-alumina powder with unstabilized zirconia ranging from 1.5 to 25 wt % and sintered at 1600°C. Lange et al.[47] report three techniques of fabricating zirconia containing yttria β″-alumina composites of which the conventional mixing milling, calcination (prereaction) followed by sintering was the most successful. Green and Metcalf[49] and Green[50] report on the milling and/or sedimentation mixing of β″-alumina and prereacted Y_2O_3-ZrO_2 powders followed by slip casting. Binner et al.[48] fabricated materials by *in situ*

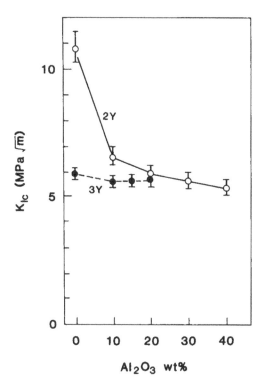

FIGURE 19.   Fracture toughness (K_{IC}) dependence upon alumina content of ZrO_2 containing 2 and 3 mol % Y_2O_3. (After Tsukuma, K., Ueda, K., Matsushita, K., and Shimada, M., *J. Am. Ceram. Soc.*, 68, C-56, 1985; Tsukuma, K., Ueda, K., Shimada, M., *J. Am. Ceram. Soc.*, 68, C-4, 1985.)

reaction of β″-alumina with sodium metazirconate during fast firing of isopressed tubes. All authors found β″-alumina was compatible with ZrO_2 with or without Y_2O_3, although in some instances β″-alumina was detected in X-ray diffraction analysis after sintering.

In the single phase β-alumina material, large grains in excess of 100 μm may be formed which act as possible sources of failure, cleaving relatively easily along their basal planes. Examples of the microstructure with or without ZrO_2 additions are shown in Figure 24. Viswanathan et al.[46] found that when the ZrO_2 was in the monoclinic form, no improvements in strength resulted. A similar effect on strength was noted by Binner et al.,[48] although these authors did find an increase in toughness with weight percent of ZrO_2 regardless of whether the phase was t- or m-ZrO_2. Green[50] found the addition of t-ZrO_2 (2.4 mol % Y_2O_3) slightly more effective than c-ZrO_2 (6.6 mol % Y_2O_3) in toughening the β″-alumina. A comparison between the various authors regarding strengths and toughness of β″-Al_2O_3-ZrO_3 composites is shown in Table 1 for 15 vol % of ZrO_2. The variations in strength and toughness found by various authors are partially due to fabrication techniques as well as methods of measuring. The influence of volume fraction of ZrO_2 on the toughness is also available (from more than one author) for differing stabilizers as shown in Figure 25.

These results indicate that transformation toughening is not very effective for improving the toughness of β″-alumina over other possible mechanisms such as crack deflection or microcracking. It should also be remembered that these solid electrolyte systems operate at 300 to 350°C, at which temperature transformation toughening would be even less effective than at room temperature.

The strengthening increase with the addition of ZrO_2 is much greater than expected from

FIGURE 20.   SEM observations of the fracture origins in Y-TZP-Al$_2$O$_3$ composites (a) sintered (1500°C, 1 hr) and (b) HIP (1500°C) after sintering. The extremely small defect at the fracture origin explains the high strength after HIP. (After Tsukuma, K., Ueda, K., Matsushita, K., and Shimada, M., *J. Am. Ceram. Soc.*, 68, C-56, 1985; Tsukuma, K., Ueda, K., Shimada, M., *J. Am. Ceram. Soc.*, 68, C-4, 1985.)

simple fracture-mechanics arguments, assuming the flaw size remains the same.[50] This conclusion is supported by observations of the fracture origins of the pure β″-alumina and those compositions containing t- or c-ZrO$_2$ additions. The significant reduction in grain size associated with the addition of ZrO$_2$ also has the effect of reducing the critical defect size, although, if care is not exercised, coarse β″-alumina or zirconia agglomerates may still limit the strength of the composite system.

The influence of volume fraction of ZrO$_2$ on the electrical conductivity of β″-alumina is shown in Figure 26. The activation energy remains constant with up to 15 vol % of ZrO$_2$ at 5.2 Kcal/mol. At low volume fractions, zirconia acts as a dispersed, inert phase in terms of sodium ion conduction and, thus, will tend to block ionic conductivity. The increase in resistivity (Ω) with volume fraction of zirconia above 15%, is more rapid than expected for a randomly dispersed second phase. Green[50] has proposed two possible explanations based on the reduction in grain size of the β″-alumina formed with the higher volume fraction of ZrO$_2$ additions.

An alternative approach to the deliberate control of grain size of the β″-alumina by the addition of zirconia to improve the toughness has been made by Troczynski and Nicholson.[51]

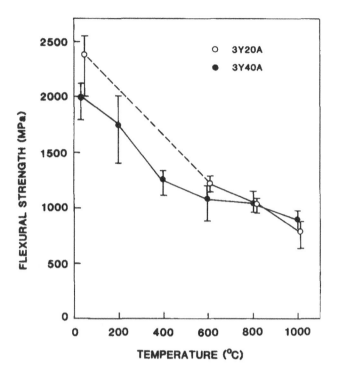

FIGURE 21. Temperature dependence of the strength of two ZrO_2-Al_2O_3 composites, 3 mol % Y_2O_3 and 20 wt % Al_2O_3 and 3 mol % Y_2O_3 and 40 wt % Al_2O_3. (After Tsukuma, K., Ueda, K., Matsushita, K., and Shimada, M., *J. Am. Ceram. Soc.*, 68, C-56, 1985; Tsukuma, K., Ueda, K., Shimada, M., *J. Am. Ceram. Soc.*, 68, C-4, 1985.)

These authors added 20 vol % of very large β″-alumina grains (~120 μm) to 8 wt % Y-PSZ (TZP), and found enhanced fracture toughness as determined from a work of fracture-type analysis. Fracture was stable during chevron-notched bend tests and the toughness increased from ≃50 J/m² for the PSZ to ≥ 500 J/m² for the composite. The strength of the composite was relatively low, ~120 MPa.

IV. SPINEL (MgO · Al_2O_3)-ZIRCONIA

Limited studies of this system do indicate that the strength of spinel (MgO · Al_2O_3) may be increased from 270 to 560 MPa with the addition of 15 vol % ZrO_2. A similar increase in K_{IC} from 2 to 4.2 MPa$\sqrt{m}$ was also observed. The optimum strengths were obtained by HIP at 1350°C and samples contained approximately 60% retained t-ZrO_2. The thermal-shock behavior of these materials indicate that the critical ΔT_c is not increased with the improved strength[26] (Figure 27).

Future directions in the spinel-zirconia system for high-temperature toughening and creep resistance appear to be possible with adjustment of stoichiometry and precipitation of Al_2O_3 in the Al_2O_3-rich spinel system. Claussen[52] has fabricated a range of compositions in the MgO-Al_2O_3 phase field with Al_2O_3 to MgO ratio increasing from one to four with and without the addition of ZrO_2. Specimens were sintered in the spinel region of the MgO · Al_2O_3 phase field as shown in Figure 28. The most promising material appeared to be Composition 3 with an Al_2O_3 · MgO ratio of two. Sintering of this composition at 1650°C with 15 vol % ZrO_2 resulted in predominantly t-ZrO_2 intergranular particles. Heat treatment of this composite at 1300°C resulted in the formation of α-Al_2O_3 precipitates in the spinel.

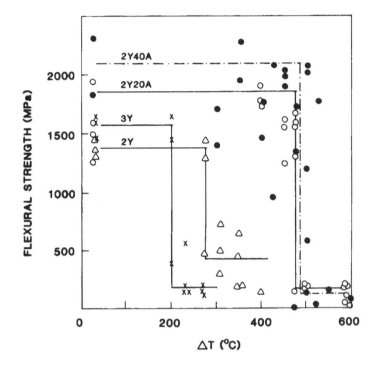

FIGURE 22. Retained strength of various ZrO_2 and ZrO_2-Al_2O_3 composites after thermal shocking into water from various temperatures. (After Tsukuma, K., Ueda, K., Matsushita, K., and Shimada, M., *J. Am. Ceram. Soc.*, 68, C-56, 1985; Tsukuma, K., Ueda, K., Shimada, M., *J. Am. Ceram. Soc.*, 68, C-4, 1985.)

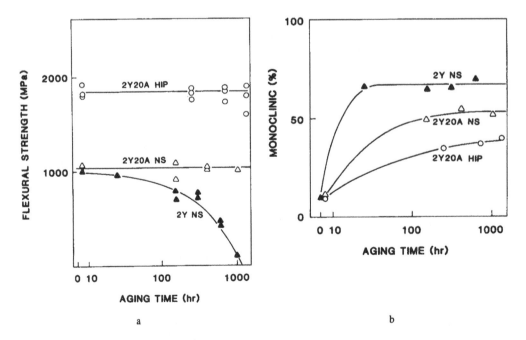

a

b

FIGURE 23. (a) The flexural strength of various ZrO_2 and ZrO_2-Al_2O_3 composites after annealing in air at 300°C for various times. (b) The development of m-ZrO_2 on the surface with time for the same compositions. (After Tsukuma, K., Ueda, K., Matsushita, K., and Shimada, M., *J. Am. Ceram. Soc.*, 68, C-56, 1985; Tsukuma, K., Ueda, K., and Shimada, M., *J. Am. Ceram. Soc.*, 68, C-4, 1985.)

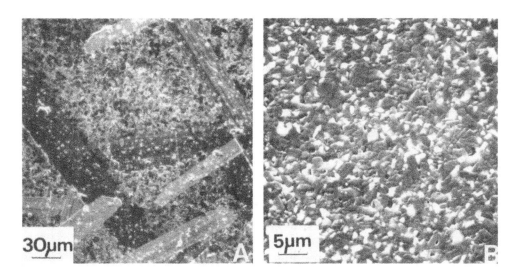

FIGURE 24. Microstructural comparison of (A) single-phase β″-Al₂O₃ and with (B) 15 vol % t-ZrO₂ composites, showing elimination of large β″ grains. (After Green, D. J., *J. Mater. Sci.*, 20, 2639, 1985.)

Table 1
COMPARISON OF MECHANICAL AND ELECTRICAL PROPERTIES OF β″-ALUMINA WITH 15 VOL% ZrO₂

ZrO₂ (phase)	K_{IC} (MPa$\sqrt{m}$)	MOR (MPa)	E (GPa)	Ω(300°C) (Ω cm)	Ref.
	2.2 — 2.6	150 — 220	202	5	Matrix only 46 — 50
t	5.0	230 — 310	—	7	46
t(2 mol% Y₂O₃)	4.5	350	—	7.7	47
t(2.4 mol% Y₂O₃)	4.1	335	—	8.5	49
c(6.6 mol% Y₂O₃)	3.2	226	—	7	50
85% m-ZrO₂	2.6	254	205		48

V. ZIRCON (ZrSiO₄)-ZIRCONIA

Garvie[53] published some evidence for improvement in the mechanical properties of the ZrSiO₄ with the addition of ZrO₂. The two phases are mutually compatible in all proportions at temperatures up to the zircon dissociation temperature, 1676°C. Zircon materials often find applications as refractories because of their relative stability to most steels at temperatures below the dissociation temperature.[54] However, generally sintering of ZrSiO₄ is difficult, and up to 10% clay is added to bond these materials.

It was found that by using plasma-dissociated zircon powders, it was possible to fabricate dense ceramics because of the reactivity of these powders. The eutectic microstructure of such powders has been analyzed by McPherson et al.[55] Bodies containing from 0 to 30 wt % ZrO₂ were prepared by standard methods and sintered at 1500°C leading to materials in excess of 90% of theoretical density. A comparison of the mechanical properties of sintered-dissociated ZrSiO₄ (DZ), DZ plus 10-wt % ZrO₂ (13 μm particle size) (DZ-10) and a clay bond zircon is shown in Table 2. The major significance of the ZrO₂ addition was the improved thermal-shock resistance which may have been predicted by the large increase in the work of fracture γ_{wof} by the addition of 10 wt % ZrO₂. The retained strength after quenching bars into water at 20°C from various temperatures is shown in Figures 29 and

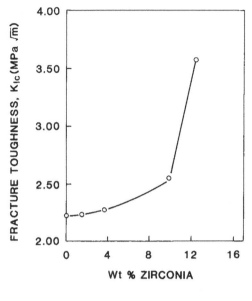

a

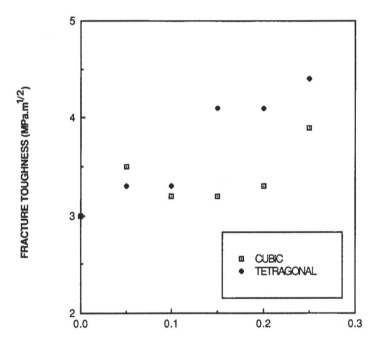

VOLUME FRACTION SECOND PHASE

b

FIGURE 25.   Dependence of the toughness (K_{IC}) of β''-Al$_2$O$_3$, (a) incorporating unstabilized ZrO$_2$ (After Binner J. G. P., Stevens, R., and Tan, S. R., in *Science and Technology of Zirconia II, Advances in Ceramics,* Vol. 12., Claussen, N., Ruhle, M., and Heuer, A. H., American Ceramic Society, Columbus, Ohio, 1984, 428.) and (b) containing both t- and c-ZrO$_2$. (After Green, D. J., *J. Mater. Sci.,* 20, 2639, 1985.)

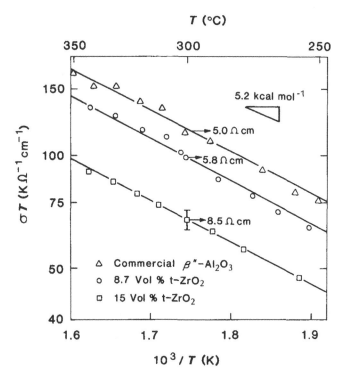

FIGURE 26. Ionic resistivity as a function of temperature for commercial β″-Al$_2$O$_3$ compared with composites containing t-ZrO$_2$. (After Green, D. J., *J. Mater. Sci.*, 20, 2639, 1985.)

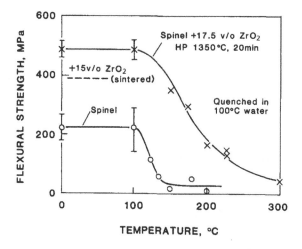

FIGURE 27. Retained flexural strength of spinel and spinel-zirconia composite after plunging into boiling water. The spinel 17 vol % ZrO$_2$ was prepared by hot pressing at 1350°C for 20 min. (After Claussen, N. and Ruhle, M., *Science and Technology of Zirconia*, 1981, 137.)

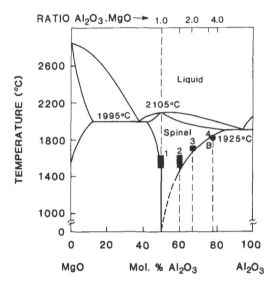

FIGURE 28.   Al$_2$O$_3$-MgO phase diagram with various spinel compositions used for precipitation-hardening by Claussen et al.[52] Heat treatment
at 1300°C led to the formation of α-Al$_2$O$_3$ precipitates only in Compositions
3 and 4.

Table 2
**MECHANICAL PROPERTIES OF ZIRCON
REFRACTORIES**

	DZ	DZ-10	Clay-bonded zircon
Flexural strength (MPa)	145 ± 15	149 ± 6	21.9
Young's Modulus (GPa)	185	188	55.6
γWOF (Jm^{-2})	19.8	73.1	20.9
γ$_f$NBT (Jm^{-2})[a]	13.0	22.2	20.8
K$_{IC}$ (MPa M$^{1/2}$)	2.2	2.9	1.5
Porosity (%)	8.1	8.0	21
Thermal expansion (°C^{-1})[b]	5.6 × 10^{-6}	5.6 × 10^{-6}	5.6 × 10^{-6}

[a]   NBT, notched bend test.
[b]   The thermal expansion data were not measured but were assumed to be the
    same for all the materials listed.

From Garvie, R. C., *J. Mater. Sci.*, 14, 817, 1979. With permission from Chapman
and Hall, publishers.

30. Bodies containing more than 10 wt % ZrO$_2$ showed no critical ΔT$_c$, but only a gradual
fall off in strength with increasing temperature.

The role of the large ZrO$_2$ particles appears to be to promote microcracking and lead to
thermal-shock cracks running through them. This is shown in Figure 31 for DZ-10 after
thermally shocking from 660°C. This figure shows that cracks in excess of 0.5 mm in length
are produced by the thermal-shock event, but do not appear to lead to significant degradation.
The large zirconia particles present in such composites and the absence of a partial stabilizer
would indicate that the zirconia was in the m-ZrO$_2$ form and that the M$_s$ temperature would
be ~1000°C. The large increase in the γ$_{wof}$ is due to the tortuous and branching nature of
the crack front as it extends through the body. This conclusion is supported by the observations of very rough fracture surface of DZ-10 material when compared to DZ.

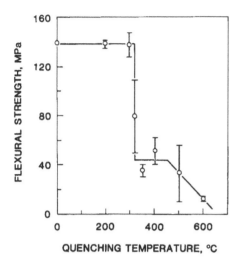

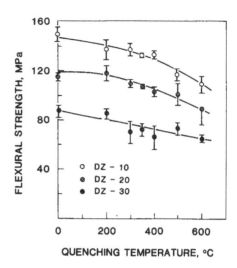

FIGURE 29.   Retained flexural strength of dissociated-zircon DZ after plunging into water at room temperature. (After Garvie, R. C., *J. Mater, Sci.*, 14, 817, 1979.)

FIGURE 30.   Retained flexural strength of dissociated zircon containing 10 (DZ-10), 20 (DZ-20), and 30 (DZ-30) wt % m-ZrO₂ after thermal shocking into water at room temperature. (After Garvie, R. C., *J. Mater. Sci.*, 14, 817, 1979.)

VI. ZINC OXIDE-ZIRCONIA

Ruf and Evans[56] found that the toughness of ZnO could be increased with the addition of up to 20 vol % ZrO_2. No mention of the influence of zirconia on strength was made. ZnO is a material with a considerably softer matrix (E = 120 GPa) than ZrO_2 (E = 200 GPa) and with a much lower yield stress or hardness (H_v = 3.6 GPa). These factors lead to considerable difficulty for the retention of the t-ZrO_2 phase without the addition of stabilizers. The lower yield point compared with other ceramics also implies that the magnitude of the residual stresses and strains about a transformed particle would be less than calculated by an elastic analysis.

Zinc oxide finds a range of applications, particularly as varistor materials. However, the mechanical properties often limit usage because the thermal shock associated with a current surge may lead to component fragmentation. Increasing the mechanical properties, particularly the toughness, has been demonstrated to greatly improve the toughness and thermal-shock resistance of zirconia-containing materials.

Ruf and Evans[56] found that in the absence of a stabilizer, almost all the zirconia (particle size range 0.1 to 1.2 μm) was m-ZrO_2 at room temperature. Addition of 4 and 8 wt % yttria to the ZrO_2 lead to retention of t-ZrO_2, particularly with the 8 wt %. They found that the hardness and Young's Modulus varied as predicted by the simple rule of mixtures. However, the toughness, K_{IC} measured by the indentation technique, was dependent upon volume fraction of ZrO_2 and the phase content. The results could be rationalized by plotting relative toughness vs. volume fraction of m-ZrO_2 as shown in Figure 32. This figure shows that the toughness increase saturates at 20 vol % ZrO_2 with a threefold increase in toughness. This increment is in good agreement with the toughness prediction on the basis of crack deflection and crack bowing mechanisms. Similar toughening increments have been reported by Swearengen et al.[57] for glass containing alumina spheres.

VII. MULLITE ($3Al_2O_3 \cdot 2SiO_2$)-ZIRCONIA

Mullite-zirconia composite systems have been studied by a number of authors.[12,58-62] In

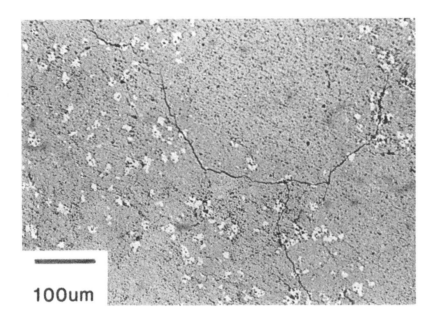

FIGURE 31. Optical micrograph of dissociated zircon — 10 wt % m-ZrO$_2$ after thermal shocking from 660°C into water. The lighter phase is m-ZrO$_2$ and the thermal shock cracks appear to intersect many of the m-ZrO$_2$ particles. (After Garvie, R. C., *J. Mater. Sci.*, 14, 817, 1979.)

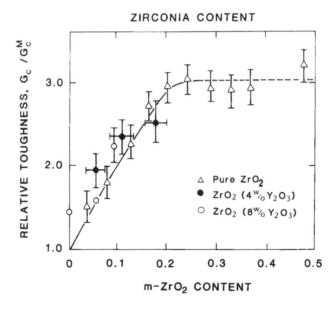

FIGURE 32. Relative critical strain energy release rate of ZnO-ZrO$_2$ composites as a function of m-ZrO$_2$ content. The K$_{IC}$ of pure ZnO is 1.7 MPa $\sqrt{m}$. (After Ruf, H. and Evans, A. G., *J. Am. Ceram. Soc.*, 66, 328, 1983.)

nearly every instance the mechanical properties are improved. Mullite, because it combines the most abundant two oxides (SiO_2 and Al_2O_3) and because of its good creep resistance at high temperature, is a potentially attractive material for high-temperature structural applications. However, the problems associated with sintering the pure material and the low toughness ($K_{IC} \sim 2$ MPa $\sqrt{m}$) and strength (~ 200 MPa) have so far limited its range of usefulness. In addition, the sintering problems usually require the addition of a glassy phase to achieve full density which degrades the high-temperature creep properties.

The addition or incorporation of ZrO_2 into mullite assists in both the sintering and an improvement of the mechanical properties. Prochazka et al.[60] found that the addition of ZrO_2 to milled-fused mullite led to lower temperature sintering than pure mullite. Apart from the ZrO_2 acting as a grain-growth inhibitor, it led to a virtual absence of glass phase along the grain boundaries. A continuous glass phase at most triple points and along some grain boundaries was observed in sintered pure mullite. The addition of 20 vol % ZrO_2 led to a virtual elimination of glass at the grain boundaries with the odd minute isolated amorphous pockets at the mullite-zirconia interface area, but not at mullite-mullite interfaces. This glass distribution most likely results from interfacial energy differences. The t-ZrO_2 content of the unstabilized zirconia addition decreased with the increase of ZrO_2 and at 20 vol % was only 16% t-ZrO_2.

The other technique for producing dense mullite-zirconia composites[61] consists of *in situ* reacting Al_2O_3 and $ZrSiO_4$ above 1450°C.

$$3Al_2O_3 + 2ZrSiO_4 \rightarrow 2ZrO_2 + 3Al_2O_3 2SiO_2$$

Further studies of this system by Claussen and Jahn[12] and Wallace et al.[58,59] indicate that materials with near-theoretical density that contained transformable t-ZrO_2 were best produced in a two-stage sinter-reaction process. Specimens were sintered at 1400°C for approximately 1 hr to 95% theoretical density, at which stage the materials were still Al_2O_3 and $ZrSiO_4$. Increasing the temperature ≥ 1500°C resulted in an *in situ* reaction to form mullite and zirconia. The reaction was virtually complete after 1 hr and resulted in a decrease in density. The major difference between *in situ*-reacted and milled-sintered materials was that much more zirconia retained the t-ZrO_2 form and was present as rounded grains in the *in situ* reacted materials. Claussen and Jahn[12] measured strengths of such materials in excess of 400 MPa, more than double that of previously reported for sintered mullite materials (150 MPa) and $K_{IC} \sim 4.5$ MPa $\sqrt{m}$. More detailed studies by Wallace et al.[58,59] indicated that the fracture toughness is a sensitive function of the transformable t-ZrO_2 content. The volume fraction of transformable ZrO_2 could be modified by changing the milling times and sintering schedules. The results of Wallace et al.[59] are shown in Figure 33 and support the concept of transformation toughening as discussed in Chapter 3.

A more detailed account of the preparation and properties of reaction-sintered mullite-ZrO_2 ceramics has been presented by Boch and Giry.[63] These authors examined the influence of various zircon-starting powders and stoichiometry, as well as sintering conditions on the properties of these composites. They find that the best properties of such *in situ* reacted materials was achieved using special zircon powders that had been plasma dissociated and calcined prior to attrition milling.

Other approaches for the fabrication of mullite-zirconia composites include sol-gel[64] and coprecipitation.[65] Both approaches provide means of fabricating high-purity starting powders to achieve bodies with a minimum of glassy phase present. Yuan et al.[64] prepared mullite powders by the sol-gel route and then blended this with very fine (≤ 0.2 μm) zirconia powders, some of which contained yttria (Y_2O_3) in order to stabilize the t-ZrO_2 phase. The sol-gel-derived powders were readily sinterable with densities of ≥ 99% of theoretical density at 1600°C. The sinterability of the coprecipitated powders was found to be strongly dependent

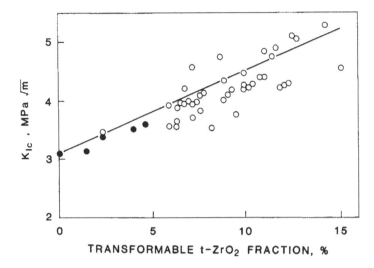

FIGURE 33.   Fracture toughness K_{Ic} various mullite-zirconia composites fabricated by *in situ* reaction as a function of transformable t-ZrO$_2$ measured on the fracture surface using X-ray diffraction. The extent of transformable t-ZrO$_2$ was modified by changing the milling times and sintering schedules. (After Wallace, J. S., Petzow, G., and Claussen, N., in *Science and Technology of Zirconia II, Advances in Ceramics,* Vol. 12, Claussen, N., Ruhle, M., and Heuer, A. H., Eds., American Ceramic Society, Columbus, Ohio, 1984, 436.)

on the stoichiometry, with pure mullite containing 74 wt % Al$_2$O$_3$ being most resistant to sintering. Densities in excess of 95% were achieved at 1700°C. The addition of as little as 10 vol % ZrO$_2$ increased the sinterability and decreased the grain size of the composite. Examples of the microstructure of these pure mullite and containing 10 vol % ZrO$_2$ are shown[65] in Figure 34.

Strength values were not determined for the gel-derived powders, although the Young's Modulus was found to decrease significantly with the addition of 20 vol % of ZrO$_2$, irrespective whether it was t- or m-ZrO$_2$. The fracture toughness was ~2.5 MPa $\sqrt{m}$, regardless of whether zirconia was added or present in the t- or m-phase. The strength of the coprecipitated-derived powders was dependent upon content of ZrO$_2$ and temperatue,[65] as shown in Figure 35. Toughness of the composite material has not been determined, nor the influence of ZrO$_2$ content on phase retention of the sintered bodies. The toughness of the matrix material was only 1.5 MPa $\sqrt{m}$, suggesting that the slight increases in strength observed in Figure 35, with the addition of zirconia, may be associated with only a modest increase in K_{IC}.

VIII. OTHER OXIDE COMPOSITES

A. Thoria (ThO$_2$)-Zirconia

Cannon and Ketcham[66] have investigated the preparation and mechanical properties of thoria-zirconia composites. Materials were prepared by milling-mixing and coprecipitation techniques followed by hot pressing or sintering. The toughness, as measured by the indentation technique, was found to depend upon volume fraction of zirconia, particle size, and porosity. Fractographic observations coupled with K_{IC} measurements suggest that transformation and microcrack toughening occurred. The ThO$_2$-ZrO$_2$ composite showed a toughness increase from 2 MPa $\sqrt{m}$ for pure ThO$_2$ to >4 MPa $\sqrt{m}$ with 15 to 20 vol % ZrO$_2$.

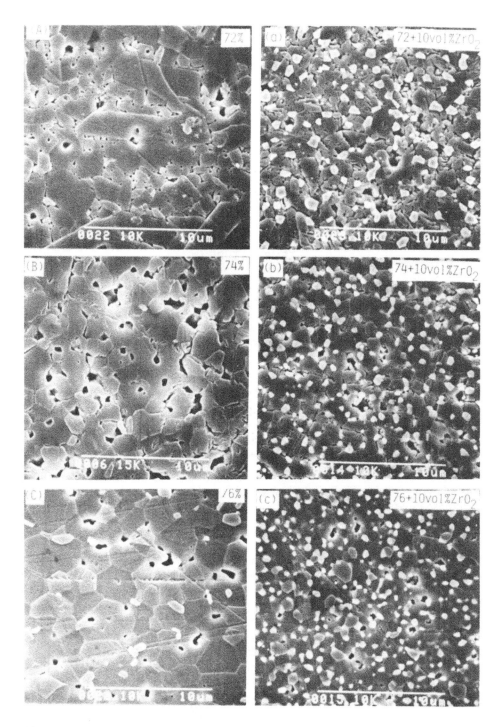

FIGURE 34. Microstructures of mullite-zirconia composites with slight variations in the $Al_2O_3 \cdot SiO_2$ stoichiometry, (a) 72%, (b) 74%, and (c) 76% and influence of 10 wt % ZrO_2 additions. (After Kubota, Y. and Takagi, H., *Special Ceramics 8*, Proceedings of the British Ceramic Society No. 37, Howlett, S. P. and Taylor, D., Eds., Institute of Ceramics, Stoke-on-Trent, England, 1986, 179.)

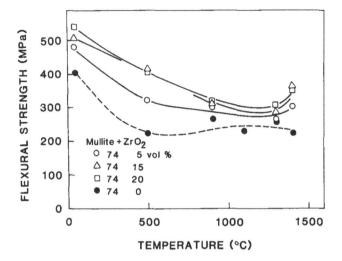

FIGURE 35. Temperature dependence of the strength of mullite zirconia composites as a function of vol % ZrO_2. All the mullite samples contained 74 wt % alumina. (After Kuabota, Y. and Takagi, H., *Special Ceramics 8*, Proceedings of the British Ceramic Society No. 37, Howlett, S. P. and Taylor, D., Eds., Institute of Ceramics, Stoke-on-Trent, England, 1986, 179.)

Table 3
PROPERTIES OF MgO-ZrO₂ COMPOSITE

Composition	Sintering temperature (°C)	Relative density (%)	Phase(s) ZrO₂	K_{Ic} MPa√m̄
Pure MgO	1300	98	—	1.8
25 wt% ZrO₂	1400	99	c-ZrO₂	2.0
25 wt% ZrO₂	1300	97	t-ZrO₂ and m-ZrO₂	3.7

From Ikuma, Y., Komatsu, W., and Yaegashi, J., *J. Mater. Sci. Lett.*, 4, 63, 1985. With permission from Chapman and Hall, publishers.

B. Magnesia (MgO)-Zirconia

Ikuma et al.[67] have reported on the toughening of MgO and a borosilicate glass by the addition of ZrO_2. They discuss a number of factors that are critical to retain ZrO_2 in the tetragonal form. These include minimal interaction of ZrO_2 with the matrix, minimal solubility of ZrO_2 in the matrix, and vice versa. Ikuma et al.[67] demonstrate the viability of this approach with MgO-ZrO_2 by sintering a coprecipitated composite to near-full density at 1300 and 1400°C. At 1400°C (as shown in Chapter 4, Figure 4) MgO and ZrO_2 react to form c-ZrO_2. This leads to a low toughness composite, whereas after sintering at 1300°C, both t- and m-ZrO_2 are present and the K_{IC} is almost double that of the matrix. The data of Ikuma et al.[67] are presented in Table 3. The operational temperature of such a composite is <1400°C which severely limits the potential refractory applications of such a composite.

C. Aluminum Titanate-Zirconia

Aluminum titanate has near-zero thermal expansion and, hence, excellent thermal-shock resistance. This arises because the thermal-expansion anisotropy stresses lead to microcracking of the material if the grain size exceeds a critical value (~10 μm). However, a con-

Aluminum Titanate

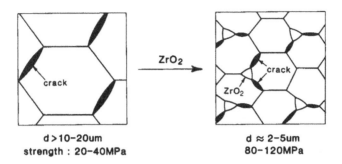

d > 10-20um
strength : 20-40MPa

d ≈ 2-5um
80-120MPa

FIGURE 36. Schematic diagram of the modification of the microstructure of $Al_2O_3.TiO_2$ by the addition of m-ZrO_2 and the resultant strengths. (After Claussen, N., in *Science and Technology of Zirconia II*, Advances in Ceramics, Vol. 12, Claussen, N., Ruhle, M., and Heuer, A. H., Eds., American Ceramic Society, Columbus, Ohio, 1984, 325.)

sequence of this excessive microcracking is that the material has low stiffness and strength (20 to 40 MPa) limiting its range of applications. At smaller grain sizes and, hence, no microcracking, the material is correspondingly stronger, but the thermal expansion becomes high 7×10^{-6} K^{-1}, and it becomes prone to thermal-shock damage.

Claussen[1] has indicated that the addition of ZrO_2 (unstabilized) enables the grain size to be reduced but still retain a low thermal expansion coefficient. This occurs because the stresses associated with the t- to m-ZrO_2 transformation are sufficient to generate microcracks which are able to absorb the large thermal-expansion anisotropy (TEA) stresses upon cooling. This situation is shown schematically in Figure 36. The reduced grain size results in improved strengths (80 to 120 MPa).

IX. SILICON NITRIDE-ZIRCONIA

Various oxide materials are added to Si_3N_4 to assist sintering and fabrication including MgO and Y_2O_3. The influence of ZrO_2 and other refractor, Zr-based materials on the mechanical properties has been examined by Rice and McDonagh,[68] and Claussen and Lahmann.[69] However, the observations by Claussen and Jahn[70] showed that attritor milling of Si_3N_4 powder with ZrO_2 and Al_2O_3 milling media (leading to a deliberate addition of Al_2O_3, up to 7.5 wt % after 6 hr) enabled both hot-pressed and sintered materials to be fabricated with improved properties. Observations of the properties of such composites, including E. Young's Modulus, K_{IC} (NB technique), flexural strength, and density vs. volume fraction of ZrO_2 are shown in Figure 37. These results indicate that the maximum strength and K_{IC} of the hot-pressed and sintered materials (with 7.5 $^w/_o$ Al_2O_3) occur with a 20 to 25 vol % loading of ZrO_2. X-Ray diffraction (XRD) analysis indicated that the silicon nitride was mainly β type, but there was up to 20 vol % of Si_2ON and that approximately 20% of the ZrO_2 had been stabilized in the cubic form by replacement of O by N in the lattice. Claussen et al.[71] have made a further study of the stabilization of ZrO_2 by mixing various metal nitrides and hot pressing in nitrogen. Most, with the notable exception of TiN, reacted to stabilize some fraction of cubic phase and to form ZrN. A detailed TEM study of the ZrO_2-ZrN system has been undertaken by Van Tendeloo and Thomas.[72] These authors found that the addition of more than 2.5 mol % ZrN to ZrO_2 lead to partial stabilization and an incommensurate modulated structure. Higher volume fractions of ZrN lead to the formation of a rhombohedral structure containing t- and m-ZrO_2 precipitates.

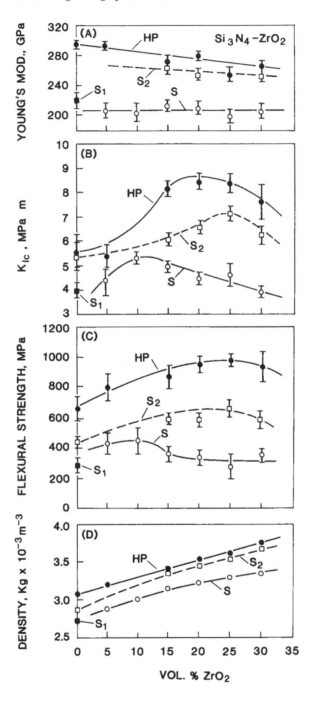

FIGURE 37.   Dependence of (A) Young's Modulus, (B) fracture toughness (K_{IC}), (C) flexural strength, and (D) density of Si_3N_4-ZrO_2 composites fabricated by hot pressing (HP) and sintering (S). S_1 and S_2 refer to specimens sintered from low-oxygen Si_3N_4 powder milled for 2 and 6 hr respectively with alumina media (6 hr of milling added 7.5 wt% of Al_2O_3). (After Claussen, N. and Jahn, J., *J. Am. Ceram. Soc.*, 61, 94, 1978.)

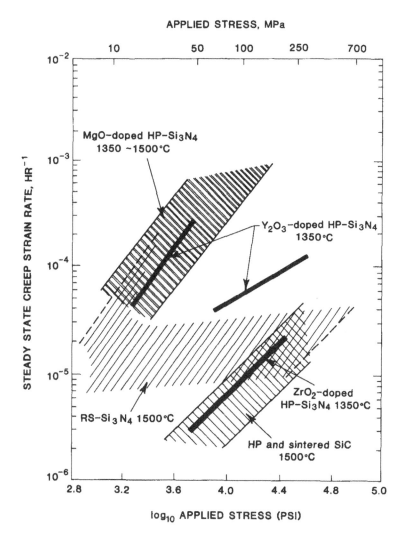

FIGURE 38.   Influence of the addition of ZrO_2 ($?Y_2O_3$) on the flexural creep behavior
of hot pressed silicon nitride materials at 1350°C. These results are compared with other
sintering additives and reaction sintered (RS) silicon nitride and silicon carbide. (After
Larsen, D. C. and Adams, J. W., Report, AFWAL-TR-83-4141 Report, Air Force Wright
Aeronautical Laboratories, Wright-Patterson Air Force Base, Ohio, April 1984.)

Other studies of the addition of ZrO_2 to silicon nitride have shown that it leads to increased
creep resistance at 1350°C. Larsen and Adams[73] have investigated the mechanical behavior
of hot-pressed Si_3N_4 containing 10% ZrO_2 (plus an unstated quantity of Y_2O_3) fabricated by
the National Aeronautics and Space Administration (NASA) Lewis Research Center, based
on a concept developed by Vasilos et al.[74] The strength of this material was somewhat less
than other hot-pressed silicon nitride materials, but even at 1500°C, it displayed no subcritical
crack growth. The stress-strain behavior of this material was linear indicating the presence
of a deformation-resistant grain boundary phase. Time to failure tests under static stress at
1400°C showed that the material had a very low slope ($-1/n$, where n = 36) where n is
the exponent in the v = AK^n relationship describing crack velocity (v) versus stress intensity
(k). The NASA material exhibited the highest n value of all the hot-pressed materials studied
by Larsen and Adams.[73] The creep behavior of this material was exceptional. Observations
of the flexural creep data are shown in Figure 38. This figure compares various silicon

nitride materials, hot pressed with different additives, as well as reaction-sintered (RS) silicon nitride and hot-pressed and sintered SiC. The addition of ZrO_2 improved the creep resistance to values comparable to the hot-pressed and sintered SiC materials.

Recently it has been appreciated that the addition of ZrO_2 to Y_2O_3 containing the Si_3N_4 is less likely to lead to cubic stabilization of the zirconia. Lange[75] has found that the zirconia can be partially stabilized by dissolution of some of the yttria from the Si_3N_4 and lead to the retention of t-ZrO_2 in the system. It seems highly probable that further exploration of this approach in the SiYAlON system may prove advantageous.

X. TITANIUM DIBORIDE-ZIRCONIA

TiB_2 is an exceptionally hard transition metal boride with the desirable properties of high Young's modulus, oxidation resistance to 1000°C, and good thermal and electrical conductivity. Another attractive feature is the wettability of this material by molten metals, particularly aluminum and its suitability or candidacy for a cathode material for Hall cells in aluminum production. However, the problems with TiB_2 have been the difficulty of sintering this material without excessive grain growth and its fragility or low K_{IC}. Recent developments in TiB_2-powder generation by CV plasma reactions between $TiCl_4$ and BH_3 with a small quantity of CH_4 have overcome many of the sintering difficulties. The above technique leads to very fine powders 0.2 to 0.4 μm which, with the presence of C, readily sinter at 1700 to 1800°C to in excess of 99% of theoretical density with minimal grain growth.[76] This has lead to strengths as high as 600 to 700 MPa being reported with a gradual decline with increasing grain size because of the hexagonal structure and thermal expansion anisotropy of TiB_2.

Another development leading to improved toughness of TiB_2 has been achieved by Watanabe and Shobu.[77] These authors, starting with coarser TiB_2 powders (2.2 μm), find that hot pressing TiB_2-ZrO_2 composites lead to significant improvements in K_{IC} and strength over the range of 20 to 75% ZrO_2. Not only was there an improvement in mechanical properties, but the porosity decreased over the same range of compositions. Examples of the strength, toughness, porosity, and hardness of composites in this system are shown in Figures 39 and 40. A typical example of the microstructure of the composite with peak strength containing 30% ZrO_2 is seen in the SEM micrograph in Figure 41. The dark continuous phase is TiB_2 and the white phase ZrO_2. XRD observations revealed that hot pressing at 1800°C and above was necessary to stabilize 40% of the ZrO_2 in the t-ZrO_2 form. The improved strength and toughness are attributed to the enhancement of densification with ZrO_2, the formation of a solid solution, (TiZr) B_2, and the presence of t-ZrO_2 which may or may not be transformable in the stress field of a crack.

More recently Shobu et al.[78] have studied the influence of partial stabilization of ZrO_2 with Y_2O_3 on the mechanical properties of TiB_2-ZrO_2 composites. The strength, toughness, and hardness of such composites are shown in Figure 42 for materials hot pressed at 1600°C. The strengths were up to 50% higher with ZrO_2 containing 2 mol % Y_2O_3. The fracture toughness shows a monotonic increase with the addition of volume fraction of t-ZrO_2, indicating that transformation toughening is the predominant toughening mode rather than microcracking as in the TiB_2-ZrO_2 (unstabilized) data shown in Figure 39.

XI. SILICON CARBIDE-ZIRCONIA

Omori et al.[79] have recently been able to synthesize SiC-ZrO_2 composites. Previous attempts to prepare such composites led to the formation of volatile products of SiO and CO at high temperature. Omori et al. found that the addition of Al_2O_3 and La_2O_3 suppressed this reaction. These authors believe that the Al_2O_3 and La_2O_3 also stabilize the t-ZrO_2,

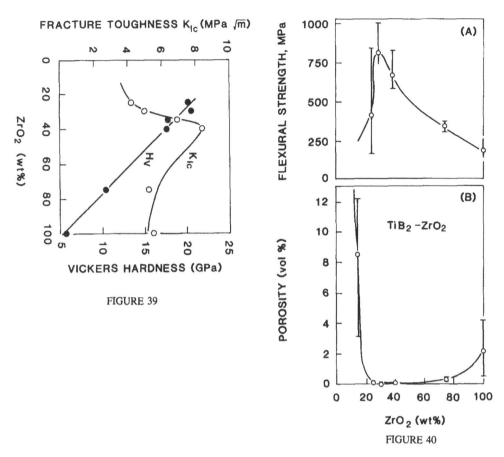

FIGURE 39

FIGURE 40

FIGURE 39   Fracture toughness (K_{IC}) and Vickers hardness of TiB_2-m-ZrO_2 composites as a function of ZrO_2 content after hot pressing at 1800°C. (After Shobu, K., Watanabe, T., Drennan, J., Hannink, R. H. J., and Swain, M. V., *Science and Technology of Zirconia III*, Advances in Ceramics, American Ceramic Society, Columbus, Ohio, in press.)

FIGURE 40.   (A) Flexural strength and (B) porosity of hot pressed TiB_2 unstabilized ZrO_2 composites with concentration of ZrO_2. (After Watanabe, T. and Shobu, K., *J. Am. Ceram. Soc.*, 68, C-34, 1985.)

although the extent of stabilization was not given. The influence of ZrO_2 content on the density and Vickers hardness is shown in Figure 43.

Claussen et al.[80-82] have investigated the incorporation of silicon carbide wishers into Y-TZP materials and Al_2O_3-m-ZrO_2 composites. The former work showed that the addition of the whiskers increased the high-temperature strength and toughness. Examples of the influence of temperature on these properties are shown in Figure 44. This figure compares Y-TZP material with and without the incorporation of whiskers.

The study of the Al_2O_3-m-ZrO_2-SiC_w composite was made in an endeavor to minimize the thermal expansion difference between Al_2O_3 and SiC_w by using the t → m-ZrO_2 transformation upon cooling.[82] It was found that the toughness of the composite increased with whisker loading and was independent of temperature to 1100°C. Figure 45 shows the variation in fracture toughness with temperature for the composite Al_2O_3-20 vol % m-ZrO_2-20 wt % SiC_w.

XII. MIXED SYSTEMS: Al_2O_3-TiC-ZrO_2

Dworak et al.[83] have examined the influence of volume fraction of these three components on the strength and toughness of composite systems. Fabrication of such mixed-based systems

FIGURE 41. SEM of TiB$_2$-30 wt % unstabilized ZrO$_2$ hot pressed at 1800°C for 30 min under a die pressure of 20 MPa in a vacuum. The darker regions are TiB$_2$ and the lighter ZrO$_2$. A phase analysis of the ZrO$_2$ revealed over half was stabilized as t- or c-ZrO$_2$.

is invariably under hot pressing conditions in a vacuum. The active phase regarding improvements in strength and toughness is invariably the zirconia which is unstabilized. Examples of the strength of similarly fabricated materials in the Al$_2$O$_3$-ZrO$_2$, Al$_2$O$_3$-TiC and Al$_2$O$_3$-TiC-ZrO$_2$ are shown in Figure 46. The peak properties in the zirconia-containing systems appear at volume loadings of 10 to 20% unstabilized zirconia. Undoubtedly further studies will take place in these mixed oxide-nonoxide-ZrO$_2$ composite systems. The availability of better-quality powders, and the desire to engineer the microstructure of ceramic materials to increase the hardness, toughness, and other properties such as electrical and thermal conductivity, will provide the basis for this work. The latter properties, particularly the electrical conductivity, enables final shaping by electrical discharge machining (EDM) which has many attractions. It may well provide the opportunity to replace strategic conventional hard materials such as WC-Co alloys.

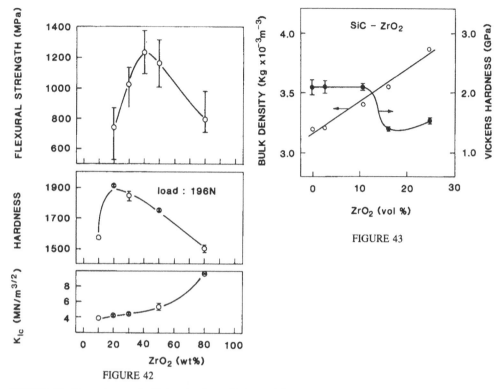

FIGURE 42   Flexural strength, Vickers hardness (20 kg) and fracture toughness (K_{IC}) of TiB_2-t-ZrO_2 (2 mol % Y_2O_3) composites hot pressed at 1600°C. (After Shobu, K., Watanabe, T., Drennan, J., Hannink, R. H. J., and Swain, M. V., *Science and Technology of Zirconia III,* Advances in Ceramics, American Ceramic Society, Columbus, Ohio, in press.)

FIGURE 43.   Density and Vickers hardness of SiC-ZrO_2 (La_2O_3-Al_2O_3) composites as a function of ZrO_2 content. (After Omori, M., Takei, H., and Ohira, K., *J. Mater. Sci. Lett.,* 4, 770, 1985.)

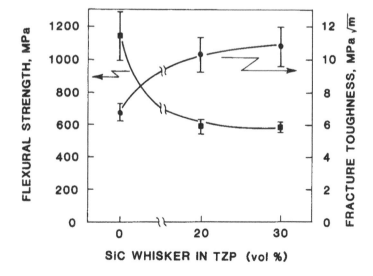

FIGURE 44.   Room temperature flexural strength and fracture toughness of Y-TZP-SiC whisker composites as a function of whisker volume fraction. (After Claussen, M., Weisskopf, K. L., and Ruhle, M., *Fracture Mechanics of Ceramics,* Vol. 7, Bradt, R. C., Evans, A. G.. Hasselman, D. P. H., and Lange, F. F., Eds., Plenum Press, 1986, 75.)

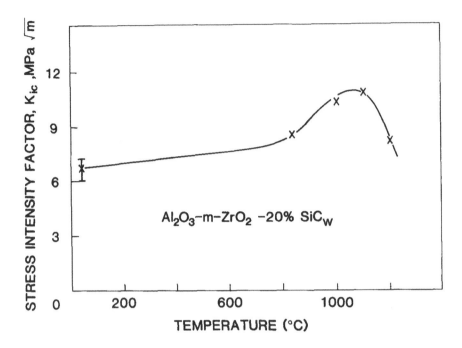

FIGURE 45.   Fracture toughness (K_{IC}) dependence upon temperature of an Al_2O_3-20 vol % m-ZrO_2 and 20 wt % SiC_w composite fabricated by hot pressing at 1500°C. (After Claussen, N., Weisskopf, K. L., and Swain, M. V., *Int. J. High Technol. Ceram.*, in press.)

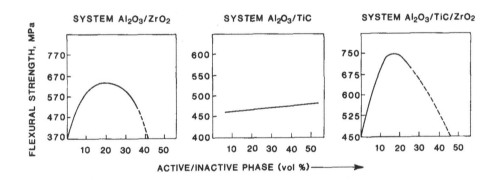

FIGURE 46.   Strength of Al_2O_3-ZrO_2 (unstabilized), Al_2O_3-TiC and Al_2O_3-TiC-ZrO_2 (unstabilized) composites as a function of ZrO_2 and TiC content. The mixed Al_2O_3-TiC-ZrO_2 composite contained equivolume fractions of Al_2O_3 and TiC. Note only with the addition of ZrO_2 do the mechanical properties pass through a maxima at 15 to 20 vol %. All specimens prepared by hot pressing. (After Dworak, U., Olapinski, H., and Thamerus, G., *Sci. Ceram.*, 9, 543, 1977.)

REFERENCES

1. **Claussen, N.,** Microstructural design of zirconia-toughened ceramics, in *Science and Technology of Zirconia, II, Advances in Ceramics,* Vol. 12, Claussen, N., Ruhle, M., and Heuer, A. H., Eds., American Ceramic Society, Columbus, Ohio, 1984, 325.
2. **Levin, E. M., Robbins, C. R., and Mc Murdie, H. F.,** Phase Diagrams for Ceramics 1964, (1969 suppl), American Ceramic Society, Columbus, Ohio, 1969.

3. **Bannister, M. J.**, Development of the SIRO$_2$ oxygen sensor: sub. solidus phase equilibria in the system ZrO$_2$-Al$_2$O$_3$-Y$_2$O$_3$, *J. Aust. Ceram. Soc.*, 18, 6, 1982.

4. **Claussen, N.**, Fracture toughness of Al$_2$O$_3$ with an unstabilized ZrO$_2$ dispersed phase, *J. Am. Ceram. Soc.*, 59, 49, 1976.

5. **Claussen, N., Steeb, J., and Pabst, R. F.**, Effect of induced microcracking on the fracture toughness of ceramics, *Am. Ceram. Soc. Bull.*, 56, 559, 1977.

6. **Kosmac, T., Swain, M. V., and Claussen, N.**, The role of tetragonal and monoclinic-ZrO$_2$ particles on the fracture toughness of Al$_2$O$_3$-ZrO$_2$ composites, *Mater. Sci. Eng.*, 71, 59, 1985.

7. **Ruhle, M., Claussen, N., and Heuer, A. H.**, Transformation and microcrack toughening as complementary processes in ZrO$_2$ toughened Al$_2$O$_3$, *J. Am. Ceram. Soc.*, 69, 195, 1986.

8. **Buresch, F. E.**, Micromechanisms controlling fracture toughness of brittle ceramics, *Powder Metall. Int.*, 12, 123, 1980.

9. **Pompe, W., Bahr, H. A., Gille, G., and Kreher, W.**, Increased fracture toughness of brittle materials by microcracking in an energy dissipative zone at the crack tip, *J. Mater. Sci.*, 13, 2720, 1978.

10. **Kreher, W. and Pompe, W.**, Increased fracture toughness of ceramics by energy dissipative mechanisms, *J. Mater. Sci.*, 16, 695, 1981.

11. **Claussen, N.**, Stress-induced transformation of tetragonal ZrO$_2$ particles in ceramic matrices, *J. Am. Ceram. Soc.*, 61, 85, 1978.

12. **Claussen, N. and Jahn, J.**, Mechanical properties of sintered, *in situ*-reacted mullite-zirconia composites, *J. Am. Ceram. Soc.*, 63, 228, 1980.

13. **Swain, M. V. and Claussen, N.**, Comparison of K_{IC} values for Al$_2$O$_3$-ZrO$_2$ composites obtained from notched bend and indentation strength measurements, *J. Am. Ceram. Soc.*, 66, C-27, 1983.

14. **Heuer, A. H., Claussen, N., Kriven, W. M., and Ruhle, M.**, Stability of tetragonal ZrO$_2$ particles in ceramics matrices, *J. Am. Ceram. Soc.*, 65, 642, 1982.

15. **Kosmac, T., Wagner, R., and Claussen, N.**, X-Ray determination of transformation depths in ceramics containing tetragonal ZrO$_2$, *J. Am. Ceram. Soc.*, 64, C-72, 1981.

16. **Claussen, N. and Petzow, G.**, Strengthening and toughening models in ceramics based on ZrO$_2$ inclusions, Proc. 4th CIMTEC, St. Vincent, Italy, 1979.

17. **Green, D. J., Lange, F. F., and James, M. R.**, Factors influencing residual surface stresses due to a stress induced phase transformation, *J. Am. Ceram. Soc.*, 66, 623, 1983.

18. **Lange, F. F.**, Transformation toughening. I—V, *J. Mater. Sci.*, 17, 225, 1982.

19. **Becher, P. F. and Tennery, V. J.**, Fracture toughness of Al$_2$O$_3$-ZrO$_2$ composites, in *Fracture Mechanics of Ceramics*, Vol. 6, Bradt, R. C., Evans, A. G., Hasselman, D. P. H., and Lange, F. F., Plenum Press, New York, 383, 1983.

20. **Becher, P. F.**, Transient thermal stress behavior in ZrO$_2$-toughened Al$_2$O$_3$, *J. Am. Ceram. Soc.*, 64, 37, 1981.

21. **Yoshimura, M., Kikugawa, S., and Somiya, S.**, *Yogyo Kyokai Shi*, 91, 182, 1983.

22. **Claussen, N., Lindeman, G., and Petzow, G.**, Rapid solidification in the Al$_2$O$_3$-ZrO$_2$ system, *Ceram. Int.*, 9, 83, 1983.

23. **Hori, S., Yoshimura, M., Somiya, S., Kunita, R., and Kaji, H.**, Mechanical properties of ZrO$_2$-toughened Al$_2$O$_3$ ceramics from CVD powders, *J. Mater. Sci. Lett.*, 1, 413, 1985.

24. **Hori, S., Yoshimura, M., and Somiya, S.**, Strength-toughness relations in sintered and isostatically hot pressed ZrO$_2$-toughened Al$_2$O$_3$, *J. Am. Ceram. Soc.*, 69, 169, 1986.

25. **Hasselman, D. P. H.**, Unified theory of thermal shock fracture initiation and crack propagation of brittle ceramics, *J. Am. Ceram. Soc.*, 52, 600, 1969.

26. **Claussen, N. and Ruhle, M.**, Design of transformation toughened ceramics, in *Science and Technology of Zirconia, Advances in Ceramics*, Vol. 3, Heuer, A. H. and Hobbs, L. W., Eds., American Ceramic Society, Columbus, Ohio, 1981, 137.

27. **Garvie, R. C.**, *Science and Technology of Zirconia III*, Advances in Ceramics, American Ceramic Society, Columbus, Ohio, in press.

28. **Green, D. J.**, Critical microstructures for microcracking in Al$_2$O$_3$-ZrO$_2$ composites, *J. Am. Ceram. Soc.*, 65, 610, 1982.

29. **Lange, F. F. and Hirlinger, M.**, Hinderance of grain growth in Al$_2$O$_3$ by ZrO$_2$ inclusions, *J. Am. Ceram. Soc.*, 67, 164, 1984.

30. **Kibbel, B. and Heuer, A. H.**, Ripening of inter and intra-granular ZrO$_2$ particles in ZrO$_2$-toughened Al$_2$O$_3$, in *Science and Technology of Zirconia II, Advances in Ceramics*, Vol. 12, Claussen, N., Ruhle, M., and Heuer, A. H., Eds., American Ceramic Society, Columbus, Ohio, 1984, 415.

31. **Kibbel, B. and Heuer, A. H.**, Exaggerated grain growth in ZrO$_2$-toughened Al$_2$O$_3$, *J. Am. Ceram. Soc.*, 69, 231, 1986.

32. **Garvie, R. C.**, Microstructure and performance of an alumina-zirconia tool bit, *J. Mater. Sci. Lett.*, 3, 315, 1984.

33. **Rieter, N.,** A new ceramic cutting material with superior toughness, *Eng. Dig. (London)*, 40, 17, 1979.
34. **Becher, P. F.,** Slow crack growth behavior in transformation toughened Al_2O_3-ZrO_2 (Y_2O_3) ceramics, *J. Am. Ceram. Soc.*, 66, 485, 1983.
35. **Lange, F. F.,** Transformation toughening, V, *J. Mater. Sci.*, 17, 255, 1982.
36. **Tsukuma, K., Ueda, K., Matsushita, K., and Shimada, M.,** High temperature strength and fracture toughness of Y_2O_3 — partially stabilized ZrO_2/Al_2O_3 composites, *J. Am. Ceram. Soc.*, 68, C-56, 1985.
37. **Tsukuma, K., Ueda, K., and Shimada, M.,** Strength and fracture toughness of hot isostatic-pressed Y_2O_3-partially stabilized ZrO_2-Al_2O_3 composites, *J. Am. Ceram. Soc.*, 68, C-4, 1985.
38. **Lange, F. F.,** Processing related fracture origins. I, *J. Am. Ceram. Soc.*, 66, 396, 1983.
38a. **Lange, F. F. and Metcalf, M. G.,** Processing related fracture origins. II, *J. Am. Ceram. Soc.*, 66, 398, 1983.
38b. **Lange, F. F., Davis, B. I., and Aksay, I. A.,** Processing related fracture origins. III, *J. Am. Ceram. Soc.*, 66, 407, 1983.
38c. **Lange, F. F., Davis, B. I., and Wright, E.,** Processing related fracture origins. IV, *J. Am. Ceram. Soc.*, 69, 66, 1986.
39. **Kummer, J. T. and Weber, N.,** A sodium-sulfur secondary battery, *S. A. E. Trans.*, 76, 1003, 1968.
40. **Weber, N.,** A thermoelectric device based on beta-alumina solid electrolyte, *Energy Convers.*, 14, 1, 1974.
41. **De Jonghe, L. C., Feldman, L. A., and Buechele, A.,** Slow degradation and electron conduction in sodium beta-aluminas, *J. Mater. Sci.*, 16, 780, 1981.
42. **Armstrong, R. D., Dickinson, T., and Turner, J.,** The breakdown of β-alumina ceramic electrolyte, *Electrochim. Acta*, 19, 187, 1974.
43. **Shetty, D. K., Virkar, A. V., and Gordon, R. S.,** Electrolyte degradation of lithia-stabilized polycrystalline β″-alumina, in *Fracture Mechanics of Ceramics*, Vol. 4, Bradt, R. C., Hasselman, D. P. H., and Lange, F. F., Eds., Plenum Press, New York, 1978, 651.
44. **Brennan, M. P. J.,** The failure of β-alumina electrolyte by a dendritic penetration mechanism, *Electrochim. Acta*, 25, 621, 1980.
45. **Virkar, A. V.,** On some aspects of the breakdown of a β″-alumina solid electrolyte, *J. Mater. Sci.*, 16, 1142, 1981.
46. **Viswanathan, L., Ikuma, Y., and Virkar, A. V.,** Transformation toughening of β″-alumina by incorporation of zirconia, *J. Mater. Sci.*, 18, 109, 1983.
47. **Lange, F. F., Davis, B. I., and Raleigh, D. O.,** Transformation strengthening of β″-Al_2O_3 with tetragonal ZrO_2, *J. Am. Ceram. Soc.*, 66, C-50, 1981.
48. **Binner, J. G. P., Stevens, R., and Tan, S. R.,** Improvement in the toughness of β″-alumina by incorporation of unstabilized zirconia particles, in *Science and Technology of Zirconia II, Advances in Ceramics*, Vol. 12, Claussen, N., Ruhle, M., and Heuer, A. H., American Ceramic Society, Columbus, Ohio, 1984, 428.
49. **Green, D. J. and Metcalf, M. G.,** Properties of slip cast transformation toughened β″-Al_2O_3/ZrO_2 composites, *Am. Ceram. Soc. Bull.*, 83, 803, 1984.
50. **Green, D. J.,** Transformation toughening and grain size control in β″-Al_2O_3-ZrO_2 composites, *J. Mater. Sci.*, 20, 2639, 1985.
51. **Troczynski, T. B. and Nicholson, P. S.,** Resistance to fracture of a partially stabilized zirconia/β-alumina composite, *J. Am. Ceram. Soc.*, 68, C-277, 1985.
52. **Claussen, N.,** Strengthening strategies for ZrO_2-toughened ceramics at high temperature, *Mater. Sci. Eng.*, 71, 23, 1985.
53. **Garvie, R. C.,** Improved thermal shock resistant refractories from plasma dissociated zircon, *J. Mater. Sci.*, 14, 817, 1979.
54. **Elderfield, R. N., Hocking, B. C., and Oxland, M. G.,** Recent advances in refractories for teeming steel, *Refract. J.*, Sept./Oct., 10, 1975.
55. **McPherson, R., Shafer, B. V., and Wong, M. M.,** Zircon-zirconia ceramics prepared from plasma dissociated zircon, *J. Am. Ceram. Soc.*, 64, C-57, 1982.
56. **Ruf, H. and Evans, A. G.,** Toughening by monoclinic zirconia, *J. Am. Ceram. Soc.*, 66, 328, 1983.
57. **Swearengen, J. C., Beauchamp, E. I., and Eagan, R. J.,** Fracture toughness of reinforced glasses, in *Fracture Mechanics of Ceramics*, Vol. 4, Bradt, R. C., Hasselman, D. P. H., and Lange, F. F., Eds., Plenum Press, New York, 1978, 973.
58. **Wallace, J. S., Petzow, G., and Claussen, N.,** Microstructure and property development in *in-situ* reacted mullite-ZrO_2 composites, in *Science and Technology of Zirconia II, Advances in Ceramics*, Vol. 12, Claussen, N., Ruhle, M., and Heuer, A. H., Eds., American Ceramic Society, Columbus, Ohio, 1984, 436.
59. **Wallace, J. S., Claussen, N., Ruhle, M., and Petzow, G.,** Development of phases in *in-situ* reacted mullite-zirconia composites, in *Surfaces and Interfaces in Ceramic and Ceramic Metal Systems*, Pask, J. A. and Evans, A. G., Eds., Plenum Press, New York, 1981.

60. **Prochazka, S., Wallace, J. S., and Claussen, N.,** Microstructure of sintered mullite-zirconia composites, *J. Am. Ceram. Soc.,* 66, C-125, 1983.

61. **Moya, J. S. and Osendi, M. I.,** Effect of ZrO_2 (ss) in mullite on the sintering and mechanical properties of mullite/ZrO_2 composites, *J. Mater. Sci. Lett.,* 2, 599, 1983.

62. **Di-Rupo, E., Carruthers, T. E., and Brook, R. J.,** Identification of stages in reactive hot pressing, *J. Am. Ceram. Soc.,* 61, 468, 1978.

62a. **DiRupo, E., Carruthers, T. E., and Brook, R. J.,** Reaction hot pressing of zircon-alumina mixtures, *J. Mater. Science,* 14, 705, 1979.

62b. **DiRupo, E., Carruthers, T. E., and Brook, R. J.,** Reaction sintering: correlation between densification and reaction, *J. Mater. Science,* 14, 2924, 1979.

62c. **DiRupo, E., Carruthers, T. E., and Brook, R. J.,** Solid state reactions in the $ZrO_2 \cdot SiO_2 \cdot \alpha \text{-} Al_2O_3$ systems, *J. Mater. Science,* 15, 114, 1980.

63. **Boch, P. and Giry, J. P.,** Preparation and properties of reaction-sintered mullite-ZrO_2 ceramics, *Mater. Sci. Eng.,* 71, 39, 1985.

64. **Yuan, Q. M., Tan, J. O., and Jin, Z. G.,** Preparation and properties of zirconia toughened mullite ceramics, *J. Am. Ceram. Soc.,* 69, 265, 1986.

65. **Kubota, Y. and Takagi, H.,** Preparation and mechanical properties of mullite and mullite-zirconia composites, *Special Ceramics,* Howlett, S. P. and Taylor, D., Eds., 1986, 179.

66. **Cannon, R. M. and Ketcham, T. D.,** Toughening of oxides by ZrO_2 additions, *Symp. on Zirconia Ceramics,* The Technical Cooperation Program meeting, U.S. Army Materials and Mechanics Research Center, Watertown, Mass., July 1979.

67. **Ikuma, Y., Komatsu, W., and Yaegashi, S.,** ZrO_2 toughened MgO and critical factors in toughening ceramic materials by incorporating zirconia, *J. Mater. Sci. Lett.,* 4, 63, 1985.

68. **Rice, R. W. and McDonagh, W. J.,** Hot pressed Si_3N_4 with Zr-based additions, *J. Am. Ceram. Soc.,* 58, 264, 1975.

69. **Claussen, N. and Lahmann, C. P.,** Fracture behaviour of some hot pressed Si_3N_4 ceramics at high temperatures, *Powder Metall. Int.,* 7, 133, 1975.

70. **Claussen, N. and Jahn, J.,** Mechanical properties of sintered and hot pressed Si_3N_4-ZrO_2 composites, *J. Am. Ceram. Soc.,* 61, 94, 1978.

71. **Claussen, N., Wagner, R., Gauckler, L. J., and Petzow, G.,** Nitride-stabilization of cubic zirconia, *J. Am. Ceram. Soc.,* 61, 369, 1978.

72. **Van Tendeloo, G. and Thomas, G.,** High resolution microscopy investigation of the system ZrO_2-ZrN, in *Science and Technology of Zirconia II, Advances in Ceramics,* Vol. 12, Claussen, N., Ruhle, M., and Heuer, A. H., American Ceramic Society, 1984, 164.

73. **Larsen, D. C. and Adams, J. W.,** Property screening and evaluation of ceramic turbine materials, AFWAL-TR-83-4141 Report, Air Force Wright Aeronautical Laboratories, Wright-Patterson Air Force Base, Ohio, April 1984.

74. **Vasilos, T., Cannon, R. M., and Wuensch, B. J.,** Improving the Stress Rupture and Creep of Silicon Nitride, National Aeronautics and Space Administration, NASA-CR-159585, Washington, D.C., 1979.

75. **Lange, F. F.,** Personal communication, 1986.

76. **Baumgartner, H. R. and Steiger, R. A.,** Sintering and properties of titanium diboride made from powder synthesized in a plasma arc heater, *J. Am. Ceram. Soc.,* 67, 207, 1984.

77. **Watanabe, T. and Shobu, K.,** Mechanical properties of hot pressed TiB_2-ZrO_2 composites, *J. Am. Ceram. Soc.,* 68, C-34, 1985.

78. **Shobu, K., Watanabe, T., Drennan, J., Hannink, R. H. J., and Swain, M. V.,** *Science and Technology of Zirconia III,* Advances in Ceramics, American Ceramic Society, Columbus, Ohio, in press.

79. **Omori, M., Takei, H., and Ohira, K.,** Synthesis of SiC-ZrO_2 composite containing t-ZrO_2, *J. Mater. Sci. Lett.,* 4, 770, 1985.

80. **Claussen, N., Weisskopf, K. L., and Ruhle, M.,** Tetragonal zirconia polycrystals reinforced with SiC whiskers, *J. Am. Ceram. Soc.,* 69, 288, 1986.

81. **Claussen, N., Weisskopf, K. L., and Ruhle, M.,** Mechanical properties of SiC whisker reinforced TZP, in *Fracture Mechanics of Ceramics,* Vol. 7, Bradt, R. C., Evans, A. G., Hasselman, D. P. H., and Lange, F. F., Eds., Plenum Press, 1986, 75.

82. **Claussen, N., Weiskopf, K. L., and Swain, M. V.,** Mechanical Properties of Al_2O_3-ZrO_2-silicon carbide whisker reinforced composites, *Int. J. High Technol. Ceram.,* in press.

83. **Dworak, U., Olapinski, H., and Thamerus, G.,** Mechanical strengthening of alumina and zirconia ceramics through the introduction of secondary phases, *Sci. Ceram.,* 9, 543, 1977.

Chapter 6

SURFACE MODIFICATION OF ZIRCONIA-TOUGHENED CERAMICS (ZTC)

I. INTRODUCTION

The reliability of a ceramic is often highly sensitive to the nature of its external surface. For example, in structural ceramics the surfaces may be subjected to high thermal or applied stresses and contact damage events. Moreover, major flaw populations, such as surface cracks, are often associated with the surface. In addition to mechanical failure processes, environmental attack often occurs preferentially at a surface. In order to overcome these difficulties, one approach is to modify the properties of the surface, such that it becomes more resistant to the particular phenomenon that is of concern. Thus, processing techniques that allow the surface of a ceramic to be modified play an important role in improving reliability in engineering applications.

There are a wide variety of surface modification techniques available and these can generally be divided into three main categories, i.e., deposition, chemical modification, and thermomechanical techniques. The aim of this chapter is to discuss those techniques that have been applied to zirconia-toughened ceramics (ZTC) and the effect that the surface modification has on the properties of the ceramic. The emphasis will be on surface modification techniques that influence the mechanical behavior of the ceramic. In this area, the aim of the modification is usually to increase the hardness or fracture toughness of the surface or to place the surface in residual compression. These three factors are important since they tend to increase the resistance to contact damage of a material. The introduction of surface compression or the increasing of the surface fracture toughness is important in that either can lead to strengthening.

The introduction of surface compression has gained most attention in ZTC because of the stress-induced phase transformation that is available in transformation-toughened materials. In the earlier part of the book, the emphasis has been on the phase transformation being induced by the stresses around a crack, but the phase transformation can also be induced at the external surface of the ceramic, which will give rise to surface compression. This is shown schematically in Figure 1. As discussed previously, the ZrO_2 phase transformation involves a volume increase as the t-ZrO_2 transforms to monoclinic. Thus, a phase transformation that occurs only at a surface will attempt to dilate the surface, but as the surface is constrained from doing so by the bulk of the material, it gives rise to surface compression.

II. MECHANICAL BEHAVIOR OF SURFACE-MODIFIED CERAMICS

It is worth considering the influence that surface modifications can have on the mechanical behavior of ceramics. The emphasis will be on increasing the hardness or fracture toughness of the surface and placing the surface under residual compression and the influence that these should have on strength and contact damage resistance.

A. Strength Behavior

Consider a through-the-thickness surface crack as shown in Figure 2, which has been introduced into a body that contains a residual surface stress distribution (σ_R). For the case in which a tensile stress, σ, is applied to the body in a direction perpendicular to the crack, the total stress intensity factor, K_I can be obtained by a superposition of the stress intensity factors due to the applied (K_I^A) and residual stress (K_I^R). These parameters can be related to the crack size and the applied and residual stresses, using a fracture mechanics approach, as introduced in Chapter 3, i.e.,

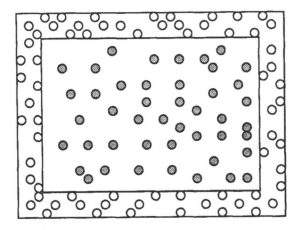

● TETRAGONAL ○ MONOCLINIC

FIGURE 1.   Schematic illustration showing ZrO_2 phase transformation occurring at the outer surface of a specimen. The volume increase associated with the transformation leads to surface compression.

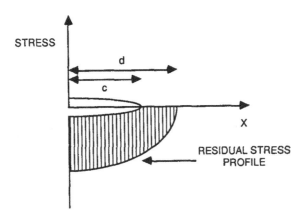

FIGURE 2.   Schematic of a surface crack (depth d) being subjected to an arbitrary residual stress distribution (depth d).

$$K_I = (K_I^A) + (K_I^R) = \sigma Y c^{1/2} + \sigma_R Y_R c^{1/2} \qquad (1)$$

where Y and Y_R are constants that depend on the crack and loading geometry and include the necessary free-surface correction terms. The critical condition for crack propagation in a brittle material is given when $K_I = K_{IC}$ and $\sigma = \sigma_f$, (See Chapter 3, Section III), i.e.,

$$K_{IC} = \sigma_f Y c^{1/2} + \sigma_R Y_R c^{1/2} \qquad (2)$$

In terms of the increase in strength, one can use Equation 1 for the case in which $\sigma_R = 0$ to obtain $K_{IC} = \sigma_f^0 Y c^{1/2}$, where σ_f^0 is the strength in the absence of residual compression. Substituting this latter expression into Equation 2, one obtains

$$\Delta\sigma_f = (\sigma_f - \sigma_f^0) = -\sigma_R(Y_R/Y) \qquad (3)$$

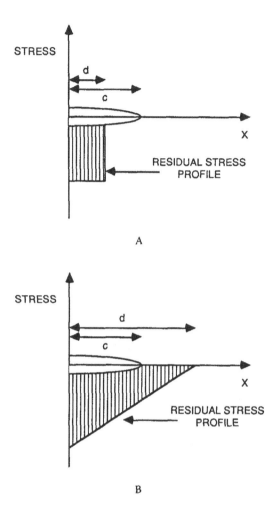

FIGURE 3. Examples of residual surface stress distributions showing (A) uniform surface compression, and (B) linear compressive surface stress gradient.

The value of Y is available in handbooks for a variety of geometries and for the case discussed here, $Y = 1.12\pi^{1/2}$. The value of Y_R is more difficult to obtain since it depends on the form of the residual stress distribution. It is, however, relatively straightforward to use a superposition and Green's Function approach to determine this parameter. For example, let us consider the case in which the surface stress represents uniform compression (σ_c) to a depth d, and the compensating interior tension (σ_T) is small enough to be neglected. This is depicted in Figure 3A. The value of Y_R is available in the literature and is given by[1]

$$Y_R = (2/\pi)(F[d/c]) \sin^{-1}(d/c) \qquad (4)$$

where F[d/c] is a free surface correction term.[1,2] For cases in which $c \leq d$, $Y = Y_R$ and $\Delta\sigma_f = -\sigma_c$, that is, the strengthening is simply equivalent to the magnitude of the surface compression. For cases in which the interior residual tensile stress is significant, an extra term must be added to Equation 1, which is given by $K_I^R = \sigma_T c^{1/2}\{(2/\pi)(G[d/c]) \cos^{-1}(d/c)\}$ and G[d/c] is a free surface correction term.[2] This type of approach has also been taken by Swain[3] for the case of a residual stress distribution in which the compressive part is linear to a depth d, while the tension is uniform in the interior of a specimen. This residual stress configuration is shown in Figure 3B. The value of Y_R for this situation had been derived previously by Lawn and Marshall.[4]

Although the above analysis was relatively straightforward, there were several simplifications made that warrant further discussion. First, although the presence of the surface compression may inhibit the extension of surface cracks, it may increase the likelihood of failure from some other flaw population. Thus, the strengthening would be less than anticipated from the above analysis. Second, the above analysis assumes that the surface crack will be completely open at failure. Compressive surface stresses would tend to press the surfaces of a crack together and, thus, a crack could go through a stage in which it is partly closed. This partial closure influences the stress intensity factor associated with a crack and, hence, the amount of strengthening. Indeed, it has been shown[5,6] that it will tend to reduce the amount of strengthening from residual surface stresses. The third important area concerns the crack shape. The above analysis was concerned with through-the-thickness cracks, whereas one expects surface cracks to have a finite depth along the surface. This problem has been analyzed recently by Lawn and Fuller[7] for the case of a penny-shaped crack subjected to a uniform surface stress to a depth d, and it was shown that the stress intensity factor at the surface is given by

$$K_I^R = M\sigma_c d^{1/2}[2 - (d/c)^{1/2}] \qquad \text{for} \quad (c \geqslant d)$$

and

$$K_I^R = M\sigma_c c^{1/2} \qquad\qquad \text{for} \quad (c \leqslant d) \qquad (5)$$

where M is a constant $\sim 2/(\pi)^{1/2}$ that also takes into account the free surface correction. The analysis by Lawn and Fuller[7] can be used for various residual stress distributions and also to determine the stress intensity factor at the deepest point of the crack. For example, the stress intensity factor at the crack depth, for the case of uniform surface compression, is given by[8]

$$K_I^R = M\sigma_c c^{1/2}\{[2 + 2(d/c)]^{1/2} - [2 - 2(d/c)]^{1/2} - (d/c)\} \qquad \text{for} \quad (c \geqslant d)$$

and

$$K_I^R = M\sigma_c c^{1/2} \qquad\qquad\qquad \text{for} \quad (c \leqslant d) \qquad (6)$$

The difference in the stress intensity factors given by Equations 5 and 6 implies the crack shape may be influenced by the surface stresses and the failure criterion becomes less well understood. If, to a first approximation, one assumed that the crack shape does remain semicircular and is controlled by the stress intensity factor at the surface, the strengthening due to uniform surface compression can be obtained from Equation 5. Ignoring the surface free correction, one obtains the following

$$\Delta\sigma_f = -\sigma_c\{(d/c)^{1/2}(2 - (d/c)^{1/2})\} \qquad (7)$$

when $c \geqslant d$ and $\Delta\sigma_f = \sigma_c$ when $c \leqslant d$.

The final area of concern in the strengthening analysis concerns the conditions under which the surface crack was first formed. As reviewed recently by Lawn,[9] surface crack formation is often a result of contact damage events and the subsequent behavior of the crack can be controlled by residual stresses at the contact site. These stresses arise as a result of the inelastic deformation at the contact point and lead to another stress intensity factor term. The effect of contact damage will be discussed in the following section.

It was indicated earlier that surface modification techniques that increase the fracture toughness (K_{IC}^S) or hardness (H) of the surface would also be a useful attribute. For surfaces with an increased fracture toughness, the modified layer depth should be greater than the crack depth to be most beneficial. Replacing K_{IC} in Equation 2 by K_{IC}^S, the strengthening with residual surface stresses (c < d) is given by

$$\Delta\sigma_F = [K_{IC}^S/K_{IC}](\sigma_F^0 - \sigma_R(Y_R/Y)) - \sigma_F^0 \qquad (8)$$

Changes in surface hardness are not expected to influence strength unless the crack is a result of a contact damage process. This latter case will be discussed in the next section and we will see that the size of cracks formed at a contact site can be reduced by an increased surface hardness.

It was indicated earlier that surface compression in ZTC can be introduced through the phase transformation. It is *important* to note in these cases, that one will also automatically *reduce the fracture toughness and hardness* of the surface. The reduced toughness is the result of not being able to stress induce the phase transformation, thereby losing transformation toughening, and the reduced hardness is because the hardness of m-ZrO_2 is less than t-ZrO_2.[10-12]

B. Contact Damage Behavior

Contact damage is important for brittle materials in that it can lead to the formation of cracks in the surface of a material. In general, there are two distinct types of cracks, i.e., lateral and radial cracks. The radial cracks form perpendicular to the external surface and radially from the contact site. The formation of these cracks is important in that it generally leads to strength degradation. The lateral cracks originate from below the contact site and are parallel to the surface initially, but can extend to the surface. The lateral cracks are generally involved in erosion and spalling types of wear.

In recent years, it has been found that hard, sharp indentors can simulate these contact events.[9] Thus, indentors can be used to introduce cracks that simulate contact-induced cracks or conversely, as a probe for studying the resistance of the surface of a material to fracture processes. For materials in which the surface is expected to be similar to the bulk, indentation can be used to measure many of the key properties of a material.[9] For this discussion, let us consider the introduction of indentation cracks into a modified surface.

First, let us consider the influence of residual surface stresses on the length of radial indentation cracks. If we consider a brittle material which is being subjected to uniform surface compression and into which we introduce an indentation, providing the load is sufficient, radial cracks will form at the indentation site. It should be pointed out that in some cases, such as ion-exchanged glass, sub-surface cracks can be formed at an indentation site below the surface compressive zone.[8] The radial cracks are generally assumed to arrest when $K_I = K_{IC}$. For a stress-free surface, this approach has been used as a technique for measuring K_{IC}.[13,14] For example, it has been shown for several ceramics that[14]

$$K_{IC} = 0.016(E/H)^{1/2}P/c_0^{3/2} \qquad (9)$$

where P is the indentation load and c_0 is the radius of the radial cracks (assumed semicircular). Equation 9 shows the importance of K_{IC} in reducing radial crack size in contact events. It is clear that the use of Equation 9 for measuring K_{IC} from indentation crack sizes presents some particular problems in ZTC. For example, as to be discussed later in this section, any phase transformation on the surface will influence c_0. In some cases, this may be difficult to prevent and the measured toughness will be higher than its real value as a result of the surface compression. In a study concerning Y-TZP, it was found that surface preparation

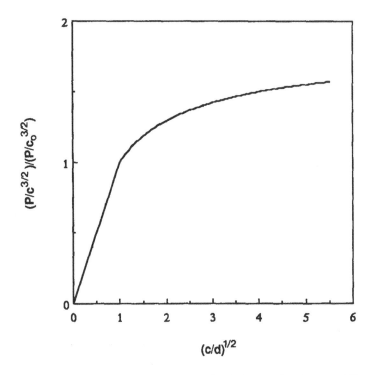

FIGURE 4. Variation of the parameter $(P/c^{3/2})/(P/c_0^{3/2})$ as a function of $(c/d)^{1/2}$ for a compressively-stressed surface, assuming a fixed positive value for $(-\sigma_c d^{1/2}/K_{IC})$.

could significantly influence the results of an indentation fracture toughness measurement.[15] As indicated earlier, the phase transformation is also expected to decrease the toughness and hardness of the surface. It has been shown that even in a nontransformed surface, a transformation zone exists around indentation sites in ZTC and this leads to an increase in the difficulty of nucleating radial cracks.[11]

For a residually stressed surface, Equation 9 needs to be modified since K_I must include contributions from both the residual stress on the surface (K_I^R) and the residual stress at the indentation site (K_I^I). This latter factor is known to depend on the indentation load,[13,14] and using this analysis, one obtains

$$K_{IC} = (K_I^R) + (K_I^I) = \sigma_c Y_R c^{1/2} + \chi P/c^{3/2} \qquad (10)$$

where χ is a constant that depends on the material parameters, Young's modulus (E) and hardness (H). Using Equation 9, the last equation can be rearranged to give

$$(P/c^{3/2}/P/c_0^{32}) = 1 - \{(\sigma_c Y_R c^{1/2})/K_{IC})\} \qquad (11)$$

Thus, one finds for the case of surface compression that the radial indentation crack size is reduced when compared to a stress-free surface. Figure 4 shows the prediction from Equation 11 using the analysis of Lawn and Fuller[7] for a penny-shaped crack (Equation 5). For a stress-free surface, the parameter $P/c^{3/2}$ should be independent of crack length or indentation load, whereas for residually stressed surfaces, $P/c^{3/2}$ should increase with $(c/d)^{1/2}$. The $P/c^{3/2}$ behavior is linear for $(c/d)^{1/2} < 1$ and this has been observed in both ion exchanged and tempered glasses.[8,16] For values of $(c/d)^{1/2} > 1$, the behavior is expected to be more complex than indicated here since the indentation cracks may no longer be considered semicircular, as discussed earlier. For example, in surface-stressed Al_2O_3-ZrO_2 composites, it has been found that $P/c^{3/2}$ approaches the value of a stress-free surface at large indentation loads.[15,17]

In addition to radial cracks being shorter on residually compressed surfaces, one would also expect the strength reduction to be less. Following the analysis of Anstis et al.,[14] the strength of indented, surface stress-free material (σ_f^0) is approximately given by

$$\sigma_f^0 = 2K_{IC}^{4/3}(H/E)^{1/6}P^{-1/3} \tag{12}$$

This equation has been used as a means of measuring K_{IC}.[14] In the same way as the measurement of indentation crack size, discussed earlier, the use of Equation 12 in measuring K_{IC} will present some special difficulties in ZTC. For a material that contains a compressively-stressed surface, the value of (σ_f^0) will be increased by the surface compression. Indeed, for the simple case of uniform compression to a depth greater than the depth of the crack, (σ_f^0) should be increased simply by $-\sigma_C$. Thus, one comes to the conclusion that compressive surface stresses can decrease the strength reduction associated with contact damage events. It is also clear from Equations 10 and 12, that increasing the fracture toughness of the surface to a depth greater than the surface crack size will be beneficial. Similar conclusions are also drawn if one considers increasing the hardness of a surface. For example, it is evident in Equations 9 and 12, that increases in H will reduce c_0 and increase σ_f^0.

It is expected that increasing surface fracture toughness and hardness will also increase the resistance to lateral crack growth at contact sites, but as pointed out recently, surface compression is expected to increase the likelihood of spalling and erosion.[18-20] Thus, one may have to use caution in using surface compression in some applications.

III. SURFACE MODIFICATION TECHNIQUES

The techniques that have been used to modify the surface of ZTC will be discussed and the initial emphasis will be on those techniques in which the aim is to induce the phase transformation at the surface. The thermodynamics involved in a constrained phase transformation were discussed in Chapters 2 and 3. By understanding those factors that are important in constraining the phase transformation, it is possible to manipulate the same factors to "unconstrain" the phase transformation. For example, directed mechanical stresses, increasing precipitate or grain size, decreasing temperature, and decreasing stabilizer additions should all act to unconstrain the phase transformation. Provided this occurs only at the external surface of the material, one should then induce surface compression. Claussen[21,22] has suggested that these processes can be classified into four groups, depending on whether the transformation is induced by stress, chemical processes that lead to the formation of m-ZrO_2, quenching below M_s, or producing a material in which M_s is above room temperature at the surface but below in the interior.

A. Mechanical Stressing

In the initial work of Garvie et al.[23] on transformation toughening, it was noticed that surface grinding was sufficient to induce the t- to m-ZrO_2 phase transformation. This idea was amplified later by Pascoe and Garvie,[24] when they showed that removal of the ground surface by polishing of a partially stabilized zirconia (PSZ) material led to a decrease in strength. The data from this investigation is shown in Figure 5. This data was subsequently analyzed by Swain[3] using an approach similar to that outlined in Section II.A. In this work, the residual stress profile was estimated from the variation of the m-ZrO_2 content at the free surface. Using the fractional change in volume associated with the phase transformation, one can determine the residual stress and then predict the strength increase as the compressive

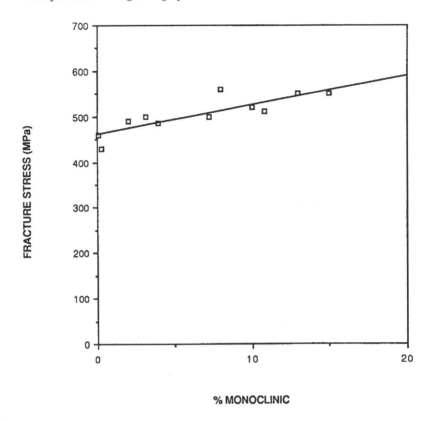

FIGURE 5.   Dependence of flexural strength on surface monoclinic content of a Ca-PSZ alloy. The decrease in monoclinic content was produced by surface polishing. (After Pascoe, R. T. and Garvie, R. C., *Ceramic Microstructures*, '76, 1977, 774.)

layer is removed by polishing. It was found that the strengthening observed by Pascoe and Garvie[24] was in excellent agreement with this analysis.[3] A later study by Reed and Lejus[25] showed that surface grinding could lead to the formation of m-ZrO$_2$ in an Y-PSZ and that there was an increase in hardness for penetration depths of ~4 μm.

The influence of surface grinding has also been studied by Gupta[26] on a tetragonal zirconia polycrystal (TZP) material. In this work, he found that the strength of the material passed through a maximum as one increases the SiC grit size of the grinding wheel. The strength increase was considered to be a result of an increase in the depth of the compressive zone with increasing grit size, whereas the decrease at the larger grit sizes was found to be associated with the formation of surface microcracks. In hot-pressed ZrO$_2$/Al$_2$O$_3$ composites, the strength of ground specimens was found to be higher than that of specimens that were ground and subsequently annealed.[27,28] Figure 6 shows an example of such data from the work of Green et al.[28]

The impact of particles on a ZTC surface can also lead to the formation of m-ZrO$_2$[29,30] and it has been shown that this can lead to an increase in strength.[28] Figure 7 shows that the strength passes through a maximum as one increases the size of the impacting particles. Presumably, at the larger grit sizes, the cracks formed by the impact are leading to a larger strength decrease than the strengthening afforded by the compressive surface. The amount of transformation was found to increase with increasing grit size. The data in Figure 7 is compared to that for specimens that were surface ground and it was determined that surface grinding was more effective as a strengthening process. The emphasis in this section has been on localized mechanical stresses, but in some cases, the phase transformation can be

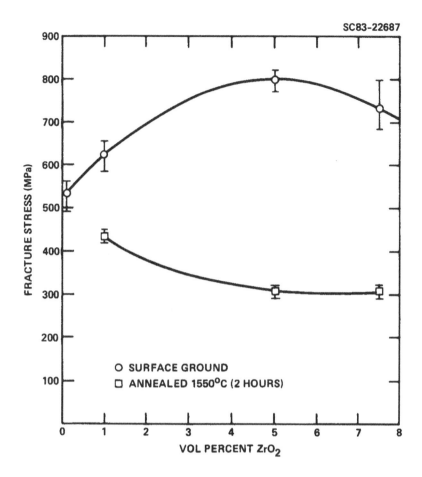

SC83-22687

FIGURE 6. Comparison of flexural strength for hot-pressed Al_2O_3-ZrO_2 composites after surface grinding with similar specimens that were annealed. (From Green, D. J., Lange, F. F., and James, M. R., *Advances in Ceramics*, Vol. 12, Science and Technology of Zirconia II, Claussen, N., Ruhle, M., and Heuer, A. H., Eds., American Ceramic Society, Columbus, Ohio, 1984, 240. With permission.)

induced by bending[31] or by thermal stress.[21] In the former study on a high-toughness Mg-PSZ, it was found that surface grinding did not influence strength and was considered to be a result of R-curve behavior.[31] The influence of surface stresses on the high toughness ZTC, therefore, appears to be an area for future research.

B. Chemical Processes

The major aim of these processes is to destabilize the surface of a material that contains t-ZrO_2. This allows M_s to be increased above room temperature in the surface region and, hence, the phase transformation can proceed. This type of approach has a major advantage over the mechanical stresses techniques discussed in the last section. In the previous approaches, once a material is raised to a high temperature, the m-ZrO_2 can transform to t-ZrO_2 and when the material is cooled it will not transform back.[28] Thus, in these cases a major disadvantage is that annealing will remove the surface compression. On the other hand, destabilizing the surface will introduce surface compressive stresses which will return after annealing.

Claussen[21,22] suggested that a heat treatment in HfO_2 powder would be a possible approach. Presumably, such a heat treatment would allow the Hf ions to diffuse into the surface region. HfO_2 is soluble in ZrO_2 and has a higher temperature associated with the tetragonal/monoclinic phase transformation and, thus, the surface region is more likely to transform during

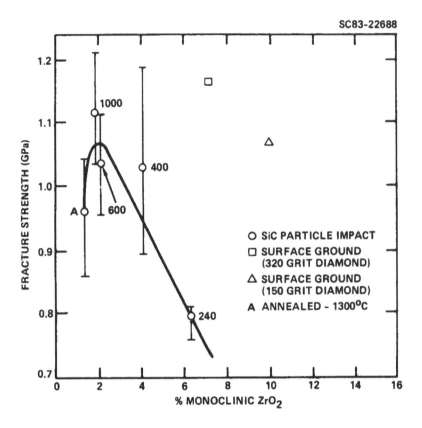

SC83-22688

FIGURE 7.   Comparison of flexural strength of an Al_2O_3 30 vol % ZrO_2 composite (2 mol % Y_2O_3) that was impacted with various sized SiC particles with similar specimens that had been surface ground. The numbers indicate the SiC grit size and the data is plotted in terms of the surface monoclinic content. (From Green, D. J., Lange, F. F., and James, M. R., *Advances in Ceramics*, Vol. 12, Science and Technology of Zirconia II, Claussen, N., Ruhle, M., and Heuer, A. H., Eds., American Ceramic Society, Columbus, Ohio, 1984, 240. With permission.)

cooling. In the same studies, Claussen[21,22] suggested that Mg-PSZ and Mg-CSZ could be simply heat treated in vacuum to induce surface stresses. The approach in this case is to allow MgO to volatilize from the surface and, hence, destabilize the surface. For materials in which the stabilizer is not volatile, a approach suggested by Green[32] could be used. In this approach, a material which needs a stabilizer to retain the t-ZrO_2, is heat treated in an unstabilized ZrO_2 powder. The stabilizer can diffuse from the surface region into the powder and when the material is cooled, the material in the denuded region can transform. This approach was demonstrated using an Al_2O_3-ZrO_2 composite, which was stabilized with Y_2O_3.[32] Figure 8 compares the apparent m-ZrO_2 content, as determined by X-ray diffraction, for a sample which underwent the Y_2O_3 removal treatment and compares the results with the profile for a ground surface. The destabilized material exhibits a larger fraction of transformed ZrO_2 in the surface region and the transformation depth is larger by a factor of ~2. The results of introducing indentation cracks into the destabilized surface are shown in Figure 9. As discussed in Section II.B, the parameter $P/c^{3/2}$ is constant for a stress-free surface, but varies with indentation load for a material that contains residual surface stress. For the materials in which the Y_2O_3 was removed, the size of the radial indentation cracks were reduced significantly when compared to the stress-free material, but the behavior is not in agreement with the trend suggested in Figure 4. The compressive surface stress in these materials was confirmed by an X-ray diffraction technique.[32]

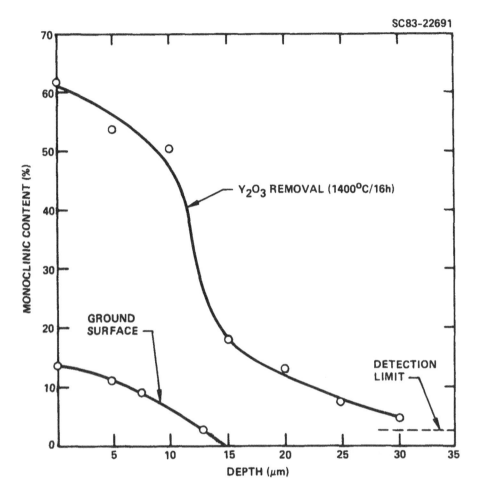

SC83-22691

FIGURE 8.   Comparison of the m-ZrO$_2$ profiles as determined by X-ray diffraction (CuKα radiation) for an Al$_2$O$_3$ 30 vol % ZrO$_2$ composite (2 mol % Y$_2$O$_3$) for the case in which the surface transformation was induced by grinding with that produced by Y$_2$O$_3$ removal. (From Green, D. J., Lange, F. F., and James, M. R., *Advances in Ceramics*, Science and Technology of Zirconia II, Claussen, N., Ruhle, M., and Heuer, A. H., Eds., American Ceramic Society, Columbus, Ohio, 1984, 240. With permission.)

C. Other Modification Techniques

Several other techniques have been suggested to induce the phase transformation at the surface and these include quenching a material for short times into liquid N$_2$ or He, increasing the amount of ZrO$_2$ in the surface region or increasing the ZrO$_2$ grain size at the surface.[21,22,33] The main difficulty with all the approaches discussed so far, is that it is difficult to produce modified surface layers that are >50 μm. Thus, in terms of strengthening, these approaches are not likely to be very effective if the material contains flaws which are much larger than this value. In response to this problem, Virkar et al.[34] have suggested an approach in which the m-ZrO$_2$ layer is introduced in the forming stage of the ceramic. Their approach is to use a "laminate" type of approach in which the outside layers are unstabilized, while the central portion is stabilized. Thus, after the material is cooled down after sintering, the outside surfaces transform. This approach has no limitations on the depth of the surface modification.

The emphasis in the discussion so far has been on materials, in which the aim is to induce the phase transformation at the surface. There are, however, a wide range of other surface modifications that could be applied to ZTC that do not depend on this phenomenon. For

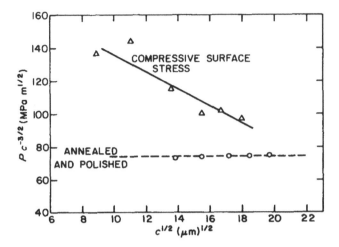

FIGURE 9.   Comparison of the variation of the parameter $P/c^{3/2}$ with the square root of indentation crack size for Al_2O_3 30 vol % ZrO_2 composites (2 mol % Y_2O_3) that had undergone the Y_2O_3 removal with that of an untreated composite. The treated specimens show a reduction in indentation crack size. (From Green, D. J., *J. Mater. Sci.*, 20, 4239, 1985. With permission from Chapman and Hall, publishers.)

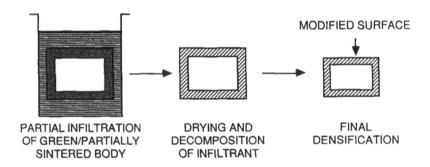

PARTIAL INFILTRATION
OF GREEN/PARTIALLY
SINTERED BODY

DRYING AND
DECOMPOSITION
OF INFILTRANT

FINAL
DENSIFICATION

FIGURE 10.   Schematic illustration of a processing technique that uses infiltration of a liquid or sol into a body before final densification. Decomposition of the inflitrant produces new phases in the surface that modify the surface properties. (From Glass, S. J. and Green, D. J., *Advances in Ceramics*, Vol. 24, Science and Technology of Zirconia III, American Ceramic Society, Columbus. Ohio, in press. With permission.)

example, in the hot pressing of Si_3N_4/ZrO_2 composites, a reaction can take place to form a zirconia oxynitride phase.[35] On oxidation, however, this phase can react to form m-ZrO_2 and the molar volume change is such that the surface is placed in compression.[36] Similar effects have been observed in sialon/ZrO_2 composites.[21]

In some situations, ZTC can undergo environmental degradation ($\sim$200°C) and one approach to overcoming this problem is to modify the surface of the material. For example, a technique was introduced by Schubert et al.[37] in which the surface was stabilized to a cubic stabilized zirconia (CSZ) and was more resistant to degradation.

A surface modification technique has been introduced, which uses partial inflitration of green or partially-sintered body by a liquid, sol, or slurry. Like the work of Virkar et al.[34] discussed earlier, this intervention earlier in the processing allows surface modification to any depth.[38] Figure 10 is a schematic illustration of the process and it has been applied to a Y-TZP material, in which the aim was to produce an Al_2O_3-ZrO_2 composite surface layer.

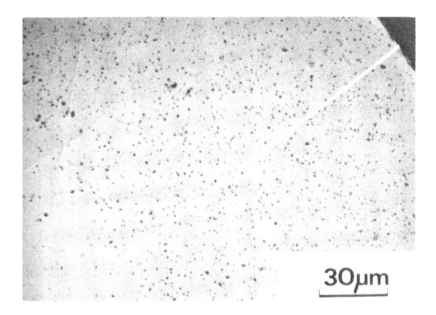

FIGURE 11. Micrograph showing the modified region in a Y-TZP in which Al_2O_3 (dark phase) was introduced into the surface region by the melt infiltration of aluminum nitrate.(From Glass, S. J. and Green, D. J., *Advances in Ceramics*, Vol. 24, Science and Technology of Zirconia III, American Ceramic Society, Columbus, Ohio, in press. With permission.)

Figure 11 shows a micrograph of the surface region for such a material and the Al_2O_3 grains with their dark contrast can be seen. The results of an indentation crack study showed that the lengths of the radial cracks were significantly reduced by the surface modification.[38]

IV. TRANSFORMATION AND SURFACE STRESSES

For those materials in which a surface stress is introduced by inducing the ZrO_2 phase transformation at a surface, it is often necessary to determine the variation of the m-ZrO_2 content in the surface region. This information is important since it can be used to estimate the residual stress profile. In the work of Pascoe and Garvie,[24] an iterative process of X-ray diffraction and polishing was used. In this type of process, the m-ZrO_2 content is not the actual value, but is an apparent value that represents the mean concentration of the zone penetrated by the X-rays. Figure 12 shows some examples of such apparent m-ZrO_2 profiles for ground surfaces of Al_2O_3/ZrO_2 composites.[39] This approach allows the transformation depth to be estimated from the depth at which the m-ZrO_2 content is the same as the bulk m-ZrO_2 content. It was found that the transformation depth depends on both the volume fraction of ZrO_2 and the amount of stabilizer (Y_2O_3) alloyed with the ZrO_2.[39] This type of polishing approach tends to be laborious and, in addition, one needs to manipulate the data further to obtain the true transformation profile. More rapid X-ray diffraction techniques have been introduced for estimating the transformation depth[40,41] and the true surface m-ZrO_2 content.[41] These techniques involve assumptions about the profile shape, but obviate the need for polishing. In addition, these techniques are particularly useful in estimating transformation depths on fracture surfaces. Techniques, other than X-ray diffraction, can be used for estimating the presence of m-ZrO_2. For example, Raman microprobe analysis has been used to determine the m-ZrO_2 in the vicinity of fracture surfaces[42] and could be used in a similar fashion for transformed external surfaces. The emphasis in this discussion has

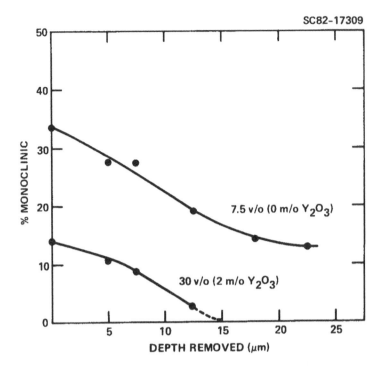

FIGURE 12. Depth of surface phase transformation in various Al₂O₃-ZrO₂ composites produced by surface grinding. The monoclinic contents were determined by X-ray diffraction on ground surface that were polished to remove the transformed layer.

been on determining the m-ZrO₂ concentration profile, but the surface residual stresses may be due to the presence of phases other than m-ZrO₂. For these cases, the approach will be similar and one would need to calculate or measure the concentration profile of the species that is giving rise to the residual stress. A further possibility is that the surface layer has a different thermal expansion coefficient than the bulk material, and in these cases, residual stresses arise when a body is cooled (or heated). Indeed, as we saw in Chapter 4, the thermal expansion of m- and t-ZrO₂ is different ($\alpha_m < \alpha_t$). Thus, in cases where the surfaces are transformed at high temperature, one would expect surface compression as a result of the expansion difference.

There are two possible approaches to determine the residual stress profiles. The first requires the concentration profile of the species that is giving rise to the dilation (or contraction) in the surface. In many situations, the residual stress profile can be determined from thermal stress theory because of the correspondence between the stresses developed by concentration gradients and the stresses developed by temperature differences. These types of approaches generally assume that the species giving rise to the residual stresses are not influencing the elastic properties of the surface. For example, Richmond et al.[43] have considered the residual stresses that arise from concentration gradients in cylindrical rods or plates. In both cases, the surface of the material is subjected to a biaxial surface stress. The maximum value of this stress occurs when the layer is thin compared to the size of the specimen and is given by

$$\sigma_{max} = -EV_v(\Delta V/V)/[3(1 - \nu)] \tag{13}$$

where E and ν are the Young's Modulus and Poisson's ratio of the surface layer, V_v is the

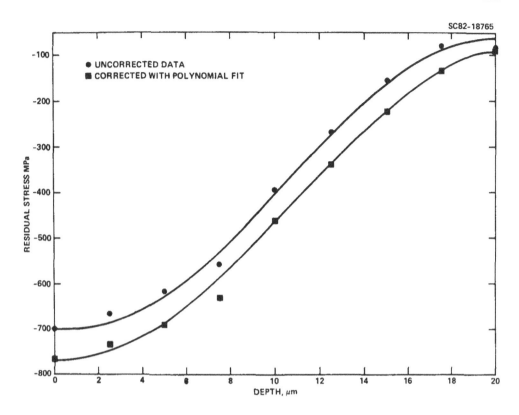

FIGURE 13. Experimentally determined surface residual stress profile for an Al₂O₃ 30 vol % ZrO₂ composite. The data were also corrected to allow for the change in profile that occurs when part of the surface is removed (by polishing) prior to each measurement. (From Green, D. J., Lange, F. F., and James, M. R., *Advances in Cermics*, Vol. 12, Science and Technology of Zirconia II, Claussen, N., Ruhle, M., and Heuer, A. H., Eds., American Ceramic Society, Columbus, Ohio, 1984, 240. With permission.)

volume fraction of the material that is giving rise to the residual stress and $(\Delta V/V)$ is the fractional volume change associated with this material. If we use values in Equation 13 that are typical of the ZrO_2 phase transformation, i.e., $(\Delta V/V) \sim 0.04$, $E \sim 200$ GPa, $V_v = 0.3$, and $\nu \sim 0.25$, a stress of -1.1 GPa is calculated, assuming no processes of stress relaxation are available. As expected, the surface is predicted to be in compression. It should be remembered that there will also be a compensating tension in the interior of the body. The maximum stress calculated above, however, is for a situation in which the surface layer is so thin when compared with the rest of the body, that the tensile stresses are very low. For situations, in which the residual stress arises because of a difference in expansion coefficient, the maximum stress is given by the same equation except the term, $V_v(\Delta V/V)/3$ is replaced by $(\Delta\alpha\Delta T)$, where $\Delta\alpha$ is the difference in the thermal expansion coefficient of the bulk and the surface and ΔT is the difference between the final temperature and the stress-free temperature.[44]

As an alternative to these calculations, there are more direct techniques available for measurement of residual stress. One of these techniques, which involves measurement of the shift in the position of an X-ray diffraction peak has been used to study the residual stresses on the ground surfaces of Al_2O_3/ZrO_2 composites.[39] Figure 13 shows an example of a residual stress profile that was obtained by this technique. The surface stresses were measured after successively removing the surface by a polishing procedure and were corrected for the stress relaxation that occurs as one removed the surface layer. Comparison with Figure 12 shows that the depth of the stresses is very similar to the transformation depth. It was also found that the magnitude of the stresses agreed reasonably well with values

Table 1

**COMPARISION OF MEASURED AND CALCULATED
RESIDUAL STRESSES IN ALUMINA/ZIRCONIA
COMPOSITES**

Vol % zirconia	Mol % yttria	Estimate vol % transformed	Surface stress (MPa)	
			Calculated	Measured
7.5	1.4	4.0	−300	−440
15	1.4	8.9	−440	−570
30	2.4	4.8 to 7.5	−310 to −480	−450 to −680
50	2.4	8.5	−480	−780
60	3.6	24.6	−1300	−1010

calculated from Equation 13. The calculated and measured values are given in Table 1 and, as can be seen, stresses as high as 1 GPa have been measured on surfaces of Al_2O_3/ZrO_2 composites.[39] Residual stresses in Al_2O_3-ZrO_2 composites have also been measured by attaching a strain gage to one surface and removing the opposite surface[34] by polishing or grinding. The amount of bending can then be related to the surface stress.

REFERENCES

1. **Tada, H., Paris, P. C., and Irwin, G. R.,** *The Stress Analysis of Cracks Handbook,* Del Research Corporation, St. Louis, 1973, p. 8.4.
2. **Hartranft, R. J. and Sih, G. C.,** Alternating method applied to edge and surface crack problems in *Mechanics of Fracture,* Vol. 1, Sih, G. C., Ed., Noordhoff International Publishing, Leyden, The Netherlands, 1973, 179.
3. **Swain, M. V.,** Grinding-induced tempering of ceramics containing metastable tetragonal zirconia, *J. Mater. Sci. Lett.,* 15, 1577, 1980.
4. **Lawn, B. R. and Marshall, D. B.,** Contact fracture resistance of physically and chemically tempered glass plates: a theoretical model, *Phys. Chem. Glasses,* 18, 7, 1977.
5. **Green, D. J.,** Compressive surface strengthening of brittle materials by a residual stress distribution, *J. Am. Ceram. Soc.,* 66, 807, 1983.
6. **Green, D. J.,** Compressive surface strengthening of brittle materials, *J. Mater. Sci.,* 19, 2165, 1984.
7. **Lawn, B. R. and Fuller, E. R., Jr.,** Measurement of thin-layer surface stresses by indentation fracture, *J. Mater. Sci.,* 19, 4061, 1984.
8. **Green, D. J.,** Unpublished results, 1986.
9. **Lawn, B. R.,** The indentation crack as a model surface flaw, in *Fracture Mechanics of Ceramics,* Vol. 5, Bradt, R. C., Evans, A. G., Hasselman, D. P. H., and Lange, F. F., Eds., Plenum Press, New York, 1983, 1.
10. **Ingel, R. P.,** Structure-mechanical property relationships for single crystal yttrium oxide stabilized zirconium oxide, Ph.D. thesis, The Catholic University, Washington, D.C., 1982.
11. **Hannink, R. H. J. and Swain, M. V.,** Induced plastic deformation of zirconia, in *Plastic Deformation of Ceramics II,* Tressler, R. E. and Bradt, R. C., Eds., Plenum Press, New York, 1984, 695.
12. **Chen, I.-W.,** Implications of transformation plasticity in ZrO_2-containing ceramics: II, Elastic-plastic indentation, *J. Am. Ceram. Soc.,* 69, 189, 1986.
13. **Lawn, B. R., Evans, A. G., and Marshall, D. B.,** Elastic/plastic indentation damage in ceramics: the median/radial crack system, *J. Am. Ceram. Soc.,* 63, 574, 1980.
14. **Anstis, G. R., Chantikul, P., Lawn, B. R., and Marshall, D. B.,** A critical evaluation of indentation techniques for measuring fracture toughness: I and II, *J. Am. Ceram. Soc.,* 64, 533, 1981.
15. **Green, D. J. and Maloney, B. R.,** Influence of surface stress on indentation cracking, *J. Am. Ceram. Soc.,* 69, 223, 1986.
16. **Marshall, D. B. and Lawn, B. R.,** An indentation technique for measuring stresses in tempered glass surfaces, *J. Am. Ceram. Soc.,* 60, 86, 1977.

17. **Green, D. J.**, Comments on "Crack-size dependence of fracture toughness in transformation-toughened ceramics", *J. Mater. Sci.*, 20, 4239, 1985.
18. **Evans, A. G. and Hutchison, J. W.**, On the mechanics of delamination and spalling in compressed films, *Int. J. Solid Struct.*, 20, 455, 1984.
19. **Marshall, D. B. and Evans, A. G.**, Measurement of adherence of residually stressed thin films by indentation, I. Mechanics of delamination, *J. Appl. Phys.*, 56, 2632, 1984.
20. **Rossington, C., Marshall, D. B., Evans, A. G., and Khuri-Yakub, B. T.**, Measurement of adherence of residually stressed thin films by indentation, II. Experiments with ZnO/Si, *J. Appl. Phys.*, 56, 2639, 1984.
21. **Claussen, N.**, Unwandlungsverstarkte Keramische Werkstoffe, *Z. Werkstofftech*, 13, 138, 1982.
22. **Claussen, N. and Ruhle, M.**, Design of transformation-toughened ceramics, in *Science and Technology of Zirconia*, Advances in Ceramics, Vol. 3, Heuer, A. H. and Hobbs, L. W., Eds., The American Ceramic Society, Columbus, Ohio, 1981, 137.
23. **Garvie, R. C., Hannink, R. H. J., and Pascoe, R. T.**, Ceramic Steel, *Nature (London)*, 258, 703, 1975.
24. **Pascoe, R. T. and Garvie, R. C.**, Surface strengthening of transformation toughened zirconia, in *Ceramic Microstructures, '76*, Fulrath, R. M. and Pask, J. A., Eds., Westview Press, Boulder, Colo., 1977, 774.
25. **Reed, J. S. and Lejus, A.**, Effect of grinding and polishing on near-surface phase transformations in zirconia, *Mater. Res. Bull.*, 12, 949, 1977.
26. **Gupta, T. K.**, Strengthening by surface damage in metastable tetragonal zirconia, *J. Am. Ceram. Soc.*, 63, 117, 1980.
27. **Claussen, N. and Petzow, G.**, Strengthening and toughening models in ceramics based on ZrO_2 inclusions, in *Energy and Ceramics*, Vincenzini, P., Ed., Elsevier Press, Amsterdam, 1980, 680.
28. **Green, D. J., Lange, F. F., and James, M. R.**, Residual surface stresses in Al_2O_3/ZrO_2 composites, in *Science and Technology of Zirconia, II*. Advances in Ceramics, Vol. 12, Claussen, N., Ruhle, M., and Heuer, A. H., Eds., The American Ceramic Society, Columbus, Ohio, 1984, 240.
29. **Lange, F. F. and Evans, A. G.**, Erosive damage depth in ceramics: a study on metastable tetragonal zirconia, *J. Am. Ceram. Soc.*, 62, 62, 1979.
30. **Wagner, R.**, The transformation of tetragonal ZrO_2 particles in an Al_2O_3 matrix and their influence on mechanical properties, Ph.D. thesis, University of Stuttgart, 1981.
31. **Marshall, D. B.**, Strength characteristics of transformation-toughened zirconia, *J. Am. Ceram. Soc.*, 69, 173, 1986.
32. **Green, D. J.**, A technique for introducing surface compression into zirconia ceramics, *J. Am. Ceram. Soc.*, 66, C178, 1983.
33. **Claussen, N. and Jahn, J.**, Transformation of ZrO_2 particles in a ceramic matrix, *Ber. Dtsch. Keram. Ges.*, 55, 487, 1978.
34. **Virkar, A. V., Huang, J. L., and Cutler, R. A.**, Strengthening of oxide ceramics by transformation induced stresses, *J. Am. Ceram. Soc.*, 70, 164, 1987.
35. **Claussen, N., Wagner, R., Gaukler, L. J., and Petzow, G.**, Nitride-stabilized cubic zirconia, *J. Am. Ceram. Soc.*, 61, 369, 1978.
36. **Lange, F. F.**, Compressive surface stresses developed in ceramics by an oxidation-induced phase change, *J. Am. Ceram. Soc.*, 63, 38, 1980.
37. **Schubert, H., Claussen, N., and Ruhle, M.**, Surface stabilization of Y-TZP, *Proc. Br. Ceram. Soc.*, 34, 1984.
38. **Glass, S. J. and Green, D. J.**, Surface modification of ceramics by partial infiltration, *Adv. Ceram. Mater.*, 2, 129, 1987.
39. **Green, D. J., Lange, F. F., and James, M. R.**, Factors influencing residual surface stresses due to a stress-induced phase transformation, *J. Am. Ceram. Soc.*, 66, 623, 1983.
40. **Kosmac, T., Wagner, R., and Claussen, N.**, X-ray determination of transformation depths in ceramics containing tetragonal ZrO_2, *J. Am. Ceram. Soc.*, 64, C72, 1981.
41. **Garvie, R. C., Hannink, R. H. J., and Swain, M. V.**, X-ray analysis of the transformed zone in partially stabilized zirconia (PSZ), *J. Mater. Sci. Lett.*, 1, 437, 1982.
42. **Clarke, D. R. and Adar, F.**, Measurement of the crystallographically transformed zone produced by fracture in ceramics containing tetragonal zirconia, *J. Am. Ceram. Soc.*, 65, 284, 1982.
43. **Richmond, O., Leslie, W. C., and Wriedt, H. A.**, Theory of residual stresses due to a chemical concentration gradient, *ASM Trans. Q.*, 57, 294, 1964.
44. **Oel, H. J. and Frechette, V. D.**, Stress distribution in multiphase systems: I, Composites with planar interfaces, *J. Am. Ceram. Soc.*, 50, 542, 1967.

Appendix

SELECTED PROPERTIES OF ZIRCONIA AND ITS ALLOYS

I. INTRODUCTION

Some selected properties of zirconia and its alloys have been listed in this appendix, to serve as a preliminary database for reference purposes. Wherever possible, data on single crystals or dense polycrystalline materials have been used. The reader should observe some caution in using this data since there are many factors that can influence properties, e.g., purity, porosity, etc. It is recommended that the reader refer to the original sources to obtain more details on the experimental conditions and the types of materials used.

II. ELASTIC CONSTANTS OF SINGLE CRYSTALS

A. Room Temperature

Composition (m/o Y_2O_3)	Elastic Constants (Gpa)			Density (kg/m³)	Ref.
	c_{11}	c_{12}	c_{44}		
8.13	401	96	55.8	6014	1
11.09	403.5	102.4	59.9	5953	1
12.08	405.1	105.3	61.8	5932	1
15.52	397.6	108.6	65.8	5868	1
17.88	390.4	110.8	69.1	5829	1
2.8—12	410	110	60	5910 — 6080	2

B. Temperature Dependence[1]

1. 11.1 m/o Y_2O_3

Temperature (°C)	c_{11}	c_{12}	c_{44}
20	403.5	102.4	59.9
100	400.2	101.8	59.0
200	395.6	101.0	57.7
300	389.8	99.8	56.0
400	382.0	97.1	54.1
500	373.1	91.2	52.2
600	364.7	88.9	50.1
700	356.6	86.8	48.0

2. 12.1 m/o Y_2O_3

Temperature (°C)	c_{11}	c_{12}	c_{44}
20	405.1	105.3	61.8
100	401.6	104.8	60.9
200	396.6	104.3	59.5
300	391.2	102.9	57.9
400	385.1	100.4	56.4
500	378.9	93.4	54.8
600	372.1	86.9	53.2
700	364.8	86.8	51.5

3. 15.5 m/o Y₂O₃

Temperature (°C)	c_{11}	c_{12}	c_{44}
20	397.6	108.6	65.8
100	394.5	108.2	64.9
200	389.8	107.3	63.5
300	384.7	106.4	51.9
400	379.5	105	59.9
500	373.6	102.1	57.7
600	366.8	98.2	55.3
700	360.2	96.3	52.6

4. 17.9 m/o Y₂O₃

Temperature (°C)	c_{11}	c_{12}	c_{44}
20	390.4	110.8	69.1
100	387.0	109.9	68.1
200	382.8	108.9	66.7
300	378.1	107.5	65.2
400	373.4	105.6	63.3
500	368.2	100.5	61.2
600	361.4	98.1	58.8
700	353.8	94.0	56.2

From Kandil, H. M., Greiner, J. D., and Smith, J. F., *J. Am. Ceram. Soc.*, 67, 341, 1984. With permission.

III. ELASTIC CONSTANTS OF POLYCRYSTALS

A. Laboratory ZTC

Compositon (m/o alloying oxide)	Phase	Temperature (°C)	Young's Modulus (GPa)	Shearmodulus (GPa)	Ref.
0	m-	25	244.2	96.3	3
		1000	153	66.1	
2Y	t-	RT	207		4
3Y	t-	RT	200		5
Mg	c-	20	169		6
		300	135		
		465	127		
		570	116		
		700	115		
		850	114		
		940	114		
		1050	114		
		1180	109		
		1225	105		
		1290	101		
		1360	94.1		
8.4 Ca	c-, t-[a]	RT	210		7
	c-, m-[b]	RT	197		

[a]   As-fired (grain size ~ 65μm).
[b]   Aged 142 hr at 1300°C (grain size ~ 100μm).

B. Commercial ZTC

Material	Temperature (°C)	Young's Modulus (GPa)	Shear modulus (GPa)
Nilcra MS	20	205	77
	200	194	76
	400	183	73
	600	170	69
	800	158	66
	1000	153	64
	1100	151	61
Nilcra TS	20	205	83
	200	195	
	300	185	
	400	173	
	500	169	
	600—1100	169—165	

IV. HARDNESS OF SINGLE CRYSTALS

Composition (m/o Alloying oxide)	Temperature (°C)	Hardness (GPa)	Phases	Ref.
2.8 Y	RT	13.6	c-, t-	8
12 Y	RT	16.1	c-	8
14.8 Y	20	12—14.4[a]	c-	9
	600	5		
8.4 Ca	RT	17.1	c-, t-	8
17.9 Ca	RT	17.2	c-	8
8.1 Mg	RT	14.4	c-, t-	8
None	RT	6.6	m-	8

[a] Depends on orientation.

V. HARDNESS OF POLYCRYSTALLINE ZTC

Composition (m/o alloying oxide)	Temperature (°C)	Hardness (GPa)	Phases	Ref.
9 Mg	RT	6.5	m-	10
	RT	8.3	c-, m-	
	RT	11.5	c-, t-	
	RT	15.5	c-	
14 Mg	RT	13.5	c-	11
	RT	7.3	m-	
9 Mg (Nilcra-MS)	20	12.0	t-, c-, m-	12
	600	8	t-, c-, m-	
9 Mg (Nilcra-TS)	20	10.4	t-, c-, m-	12
	600	7	t-, c-, m-	
2 Y	RT	11.6	t-	4
3 Y	RT	12.7	t-	5
7.5 Y	RT	11.4	c-	4

VI. LATTICE PARAMETERS

A. Monoclinic ZrO₂: Space Group P2₁/c: Density (RT) 5750 kg/m³

Temperature (°C)	a (nm)	b (nm)	c (nm)	β (°)	Ref.
RT	0.51507(4)	0.52031(4)	0.53154(4)	99.194(4)	13
956	0.51882	0.52142	0.53836	98.783	14
RT[a]	0.51172	0.51770	0.53031	98.91	15
RT[b]	0.51928	0.52043	0.53617	98.80	16

Note: Additional data can be found in Chapter 2, Table 1.

[a]   m-ZrO₂ in 9.4 m/o MgO-ZrO₂ at 20°C for constrained phase in TEM foil.
[b]   12 Ce, measured at 20°C from polished and liquid nitrogen quenched surface.

B. Tetragonal ZrO₂: Space Group P4₂/nmc

Composition (m/o alloy oxide)	Temperature (°C)	a (nm)	c(nm)	Density (kg/m³)	Ref.
0	1152	0.51518	0.52724	5849[a]	14
2Y	RT	0.5095	0.5180	6080	17
3Y	RT	0.5096	0.5180	6080	17
4Y	RT	0.5098	0.5180	6050	17
3—13Y	RT	Footnote b	Footnote c	Footnote d	18, 19
2Y	RT	0.51007	0.51767	6060	20
2Y	600	0.51250	0.52084	6040	20
2Y	900	0.51364	0.52293		20
3Y	RT	0.51024	0.51745		20
3Y	600	0.51301	0.52073		20
3Y	900	0.51424	0.52214		20
5Y	RT	0.5116	0.5157		21
9.4 Mg	RT	0.5077	0.5183		21
8.4 Ca	RT	0.5094	0.5180		21
12 Ce	RT	0.5132	0.5228		22
7 Ce	RT	0.5114	0.5212		23
8.6 Ce	RT	0.5116	0.5214		23
9.5 Ce	RT	0.5118	0.5218		23
10.8 Ce	RT	0.5121	0.5219		23
11.0 Ce	RT	0.5124	0.5220		23
12.2 Ce	RT	0.5125	0.5220		23
15.8 Ce	RT	0.5125	0.5224		23

[a]   Calculated from lattice parameters.
[b] $a = 0.5060 + 0.000349X$, where X is m/o YO₁.₅.
[c] $c = 0.5195 - 0.000309X$, where X is m/o YO₁.₅.
[d]   Density = {818.435 − 137.013M/[100 + M]}/(cell volume nm³), where M is m/o Y₂O₃.

C. Cubic ZrO_2: Space Group F_{m3m}

Stabilizer	a (nm)	Density (kg/m³)	Ref.
9.4Mg	0.5080	—	21
8.4Ca	0.5132	—	21
5Y	0.5130	—	21
2—50Y	Footnote a	Footnote b	18, 19
None	0.5127	6070	21

[a] $a = 0.5104 + 0.0408M/[100 + M]$, where M is m/o Y_2O_3.

[b] Density $= \{818.435 - 137.023M/[100 + M]\}/(cell volume nm³)$, where M is m/o Y_2O_3.

VII. EXAMPLES OF X-RAY DIFFRACTION POWDER DATA

A. Monoclinic (RT) JCPDS File 13-307

d(A)	I/I_0	hkl
5.04	6	100
3.69	18	100
3.63	14	110
3.16	100	$11\bar{1}$
2.834	65	111
2.617	20	002
2.598	12	020
2.538	12	200
2.488	4	$10\bar{2}$
2.328	6	021
2.285	2	210
2.252	4	$11\bar{2}$
2.213	14	$21\bar{1}$
2.182	6	102
2.015	8	112
1.989	8	$20\bar{2}$
1.845	18	022
1.818	12	220
1.801	12	$12\bar{2}$
1.780	6	$22\bar{1}$
1.691	14	300, 202
1.656	14	013, 221
1.640	8	130
1.608	8	$31\bar{1}$, 212
1.591	4	$13\bar{1}$
1.581	4	$22\bar{2}$
1.541	10	131
1.508	6	113
1.495	10	$21\bar{3}$
1.476	6	311
1.447	4	$12\bar{3}$
1.420	5	321, 320

From Inorganic Powder Diffraction File Sets, Sets 11—15, Joint Committee on Powder Diffraction Standards, International Center for Diffraction Data, Swarthmore, Pa., 1972. With permission.

**B. Tetragonal (1250°C)
JCPDS File 24-1164**

d(A)	I/I$_0$	hkl
2.995	100	100
2.635	10	002
2.574	13	110
2.134	2	102
1.8412	23	112
1.8200	13	200
1.5821	8	103
1.5553	15	211
1.4975	4	202
1.2869	2	220
1.1940	4	213
1.1824	2	301
1.1728	2	114
1.0548	2	312

From Inorganic Powder Diffraction File Sets, Sets 29—30, Joint Committee on Powder Diffraction Standards, International Center for Diffraction Data, Swarthmore, Pa., 1987. With permission.

**C. Cubic (8 m/o Y$_2$O$_3$) (RT):
JCPDS File 30-1468**

d(A)	I/I$_0$	hkl
2.968	100	111
2.571	25	200
1.818	55	220
1.550	40	311
1.484	6	222
1.285	5	400
1.179	10	331
1.149	6	420
1.049	10	422
0.9891	6	511
0.9086	3	440

From Inorganic Powder Diffraction File Sets, Sets 23—24, Joint Committee on Powder Diffraction Standards, International Center for Diffraction Data, Swarthmore, Pa., 1984. With permission.

VIII. THERMAL EXPANSION COEFFICIENTS (UNITS 1/10^6 K)

A. Monoclinic ZrO$_2$ (Single Crystal)[24]

Temperature (°C)	a axis	b axis	c axis	β (°)
264	8.4	3.0	14.0	99.18
504	7.5	2.0	13.0	99.05
759	6.8	1.1	11.9	99.97
964	7.8	1.5	12.8	99.70
1110	8.7	1.9	13.6	99.72

B. Tetragonal ZrO$_2$ (Single Crystal)

Composition (m/o alloy oxide)	Temperature (°C)	a axis	c axis	Ref.
None	1150 — 1750	12.4	14.4	25
2 Y	RT — 600	8.22	10.54	20
2 Y	RT — 800	8.97	13.03	
3 Y	RT — 600	9.34	10.91	
3 Y	RT — 800	10.06	11.61	

C. Polycrystalline Data

Composition (m/o alloy oxide)	Phase	Temperature (°C)	Expansion coefficient (1/10⁶ K)	Ref.
14 Mg[a]	m-	RT — 1100	6.8	11
0[b]	m-	20 — 1180[c]	8.12	26
		20 — 1170[d]	8.66	
0[b]	t-	1190 — 1500[c]	9.93	26
		1190 — 1500[d]	10.04	
2 Y	t-	RT — 600	10.3	20
2 Y	t-	RT — 800	10.4	
3 Y	t-	RT — 600	10.1	
3 Y	t-	RT — 800	10.6	
14 Mg	c-	RT — 1100	11.2	11
2.8 Y	c-	20 — 1500	10.99	26
3.4 Y	c-	20 — 1500	10.68	
3.8 Y	c-	20 — 1500	10.69	
4.5 Y	c-	20 — 1500	10.92	
6.9 Y	c-	20 — 1500	10.23	
12 Y	c-	20 — 1500	11.08	

[a] Decomposed cubic.
[b] Transformed cubic single crystal.
[c] 1st cycle.
[d] 2nd cycle.

D. Commercial ZTC

Material	Temperature (°C)	Thermal expansion coefficient (1/10⁶ K)	Ref.
Feldmuhle TTZ	20 — 1500	11.03	26
NGK	20 — 1448	9.86	26
Coors ZDM	1300 — 1500	12.58	26
Coors ZDY	20 — 1500	10.45	26
Nilcra MS	20 — 450	10.2	12
	20 — 800	10.6	
Nilcra TS	20 — 450	9.4	12
	20 — 800	9.0	

IX. THERMAL DIFFUSIVITY[27]

A. Mg Stabilized

Composition (w/o)	Phases	Thermal diffusivity (mm²/sec)
2.6[a]	c, t, m	1.320
2.7	c, t	1.136
2.8	c, t, m	1.225
3.0	c, t, m	1.140
3.0	c, t, m	1.096
3.3	c, t, m	1.051
3.3	c, t, m	1.270
3.3	c, t, m	1.130
3.3	c, t, m	1.120
3.4[a]	c, t, m	1.120
3.4	c, t, m	1.150
3.5	c, t, m	1.102
3.9	c, t	0.774
5.0[b]	c	0.674
5.0[c]	m	2.397

Note: Thermal diffusivity t-ZrO_2 calculated to be 173 mm²/sec.

[a]   Falls to ~0.6 mm²/sec at 1000°C.
[b]   Approximate independent temperature to 1000°C.
[c]   Falls to ~0.85 mm²/sec at 1000°C.

B. Y Stabilized

Composition (w/o)	Phases	Thermal diffusivity (mm²/sec)
2.4	t, m	1.190[a]
3.6	t, m	1.110
4.0	c, t, m	1.100
4.5	t	1.227
5.0	c, t, m	1.100
5.3	c, t	1.044[b]
5.4	c, t	1.190
5.5	c, t	1.078
8.6	c, m	1.040
9.0	c	0.720[c]
16.8	c	0.700
16.9	c	0.890

[a]   Falls to ~0.6 mm²/sec at 1000°C.
[b]   Falls to ~0.56 mm²/sec at 1000°C.
[c]   Falls to ~0.5 mm²/sec at 1000°C.

From Hasselman, D. P. H., Johnson, L. F., Bentsen, L. D., Syed, R., Lee, H. L., and Swain, M. V., *Am. Ceram. Soc. Bull.*, 66, 799, 1987. With permission.

X. SPECIFIC HEAT (J/(kg·K)[27]

A. Mg Stabilized

| Temperature (°C) | Composition (w/o) | | | | | |
	2.6	2.7	3.0	3.4	5.0[a]	5.0[b]
25	471	470	474	481	486	479
100	519	523	524	529	536	520
200	557	562	563	568	577	561
300	552	590	580	595	609	577
400	590	610	610	610	628	638
500	618	629	645	631	653	632
600	627	636	653	639	678	637

[a]   Cubic.
[b]   Monoclinic.

B. Y Stabilized

Temperature (°C)	Composition (w/o)			
	2.4	4.0	5.3	16.9
25	467	466	460	471
100	519	512	509	519
200	557	550	546	558
300	554	580	575	582
400	595	602	591	603
500	616	630	600	616
600	639	641	604	634

From Hasselman, D. P. H., Johnson, L. F., Bentsen, L. D., Syed, R., Lee, H. L., and Swain, M. V., *Am. Ceram. Soc. Bull.*, 66, 799, 1987. With permission.

REFERENCES

1. **Kandil, H. M., Greiner, J. D., and Smith, J. F.**, Single crystal elastic constants of yttria-stabilized zirconia in the range 20° to 700°C, *J. Am. Ceram. Soc.*, 67, 341, 1984.
2. **Ingel, R. P., Lewis, D., Bender, B. A., and Rice, R. W.**, Physical, microstructural and thermomechanical properties of ZrO_2 single crystals, *Advances in Ceramics*, Vol. 12, American Ceramic Society, Columbus, Ohio, 1984, 408.
3. **Smith, C. F. and Crandall, W. B.**, Calculated high-temperature elastic constants for zero porosity monoclinic zirconia, *J. Am. Ceram. Soc.*, 47, 624, 1964.
4. **Lange, F. F.**, Transformation Toughening. III. Experimental observations in the ZrO_2-Y_2O_3 system, *J. Mater. Sci.*, 17, 247, 1982.
5. **Toya Soda Manufacturing Co., Ltd.**, Tokyo, Technical Bulletin No. Z-301.
6. **Ryshkewitch, E.**, *Oxide Ceramics*, Academic Press, Orlando, 1960, 380.
7. **Garvie, R. C., Hannink, R. H. J., and Urbani, C.**, Fracture mechanics study of a transformation toughened zirconia alloy in the CaO-ZrO_2 system, *Ceram. Int.*, 6, 19, 1980.
8. **Ingel, R. P., Lewis, D., Bender, B. A., and Rice, R. W.**, Physical, microstructural and thermomechanical properties of ZrO_2 single crystals, *Advances in Ceramics*, Vol. 12, American Ceramic Society, Columbus, Ohio, 1984, 408.
9. **Hannink, R. H. J. and Swain, M. V.**, *Plastic Deformation of Ceramics II*, Tressler, R. E. and Bradt, R. C., Eds., Plenum Press, New York, 1984, 695.
10. **Chen, I.-W.**, Implications of transformation plasticity in ZrO_2-containing ceramics. II. Elastic-plastic indentation, *J. Am. Ceram. Soc.*, 69, 189, 1986.
11. **Swain, M. V., Garvie, R. C., and Hannink, R. H. J.**, Influence of thermal decomposition on the mechanical properties of magnesia-stabilized zirconia, *J. Am. Ceram. Soc.*, 66, 358, 1983.
12. **Nilcra-PSZ**, Product information sheet, Victoria, Australia.
13. **Hann, R. E., Suitch, P. R., and Pentecost, J. L.**, Monoclinic crystal structures of ZrO_2 and HfO_2 refined from x-ray powder diffraction data, *J. Am. Ceram. Soc.*, 68, C-285, 1985.
14. **Kriven, W. M., Fraser, W. L., and Kennedy, S. W.**, The martensitic crystallography of tetragonal zirconia, *Advances in Ceramics*, Vol. 3, American Ceramic Society, Columbus, Ohio, 1981, 82.
15. **Hannink, R. H. J.**, Unpublished results.
16. **Hay, D.**, Commonwealth Scientific and Industrial Research Organization (Australia), Unpublished results.
17. **Tsukuma, K., Kubota, Y., and Tsukidate, T.**, Thermal and mechanical properties of Y_2O_3-stabilized tetragonal zirconia polycrystals, *Advances in Ceramics*, Vol. 12, American Ceramic Society, Columbus, Ohio, 1984, 382.
18. **Scott, H. G.**, Phase relations in the zirconia-yttria system, *J. Mater. Sci.*, 10, 1527, 1975.
19. **Ingel, R. P. and Lewis, D.**, Lattice parameters and density for Y_2O_3 stabilized ZrO_2, *J. Am. Ceram. Soc.*, 69, 325, 1986.
20. **Schubert, H.**, Anisotropic thermal expansion coefficients of Y_2O_3-stabilized tetragonal zirconia, *J. Am. Ceram. Soc.*, 69, 270, 1986.
21. **Hannink, R. H. J.**, Growth morphology of the tetragonal phase in partially stabilized zirconia, *J. Mater. Sci.*, 13, 2487, 1978.
22. **Hay, D.**, Private communication.

23. **Tsukuma, K. and Shimada, M.,** Strength, fracture toughness and Vicker's hardness of CeO_2-stabilized tetragonal ZrO_2 polycrystals (Ce-TZP), *J. Mater. Sci.*, 20, 1178, 1985.

24. **Grain, C. F. and Campbell, W. J.,** U.S. Bureau of Mines RI5982, College Park, Md., 1982.

25. **Lang, S. M.,** Axial thermal expansion of tetragonal ZrO_2 between 1150 and 1700°C, *J. Am. Ceram. Soc.*, 47, 641, 1964.

26. **Adams, J. W., Nakamura, H. H., Ingel, R. P., and Rice, R. W.,** Thermal expansion behavior of single crystal zirconia, *J. Am. Ceram. Soc.*, 68, C-228, 1985.

27. **Hasselman, D. P. H., Johnson, L. F., Bentsen, L. D., Syed, R., Lee, H. L., and Swain, M. V.,** Thermal diffusivity and conductivity of dense polycrystalline ZrO_2 ceramics: a survey, *Am. Ceram. Soc. Bull.*, 66, 799, 1987.

INDEX

Printed and bound by CPI Group (UK) Ltd, Croydon, CR0 4YY

22/10/2024

01777630-0010